MOLECULAR BIOLOGY
INTELLIGENCE
UNIT

The JNK Signaling Pathway

Anning Lin, Ph.D.
Ben May Institute for Cancer Research
University of Chicago
Chicago, Illinois, U.S.A.

CRC Press
Taylor & Francis Group
Boca Raton London New York

CRC Press is an imprint of the
Taylor & Francis Group, an informa business

The JNK Signaling Pathway

Molecular Biology Intelligence Unit

CRC Press
Taylor & Francis Group
6000 Broken Sound Parkway NW, Suite 300
Boca Raton, FL 33487-2742

First issued in hardback 2019

ISBN 13: 978-1-58706-120-2 (hbk)

Library of Congress Cataloging-in-Publication Data

The JNK signaling pathway / [edited by] Anning Lin.
p. ; cm. -- (Molecular biology intelligence unit)
Includes bibliographical references and index.
ISBN 1-58706-120-1
1. JNK mitogen-activated protein kinases--Inhibitors. 2. JNK mitogen-activated protein kinases. 3. Cellular signal transduction. 4. Cell death. I. Lin, Anning. II. Series: Molecular biology intelligence unit (Unnumbered).
[DNLM: 1. JNK Mitogen-Activated Protein Kinases--antagonists & inhibitors. 2. JNK Mitogen-Activated Protein Kinases--physiology. 3. Cell Death--physiology. 4. Signal Transduction--physiology. QU 141 J61 2006]
QP606.P76J55 2006
571.7'4--dc22

2005028599

About the Editor

Anning Lin is an Associate Professor at Ben May Institute for Cancer Research and the Program Leader of the Cell Signaling and Gene Regulation Program of Cancer Research Center, University of Chicago, U.S.A. His main research interests include the function of the intracellular signaling network in programmed cell death and tumorigenesis, using the c-Jun N-terminal protein kinase (JNK) and IκB kinase (IKK) signaling pathways as molecular probes. He serves on several editorial boards, including *Molecular and Cellular Biology*. He received his Ph.D. from University of Alabama at Birmingham (UAB) and did his Postdoctoral at University of California, San Diego (UCSD).

CONTENTS

EDITOR

Anning Lin
Ben May Institute for Cancer Research
University of Chicago
Chicago, Illinois, U.S.A.
Email: alin@huggins.bsd.uchicago.edu
Chapter 6

CONTRIBUTORS

Brydon L. Bennett
Inflammation Research
Experimental Therapeutics Group
Celgene
San Diego, California, U.S.A.
Email: bbennett@celgene.com
Chapter 8

Ewen Gallagher
Laboratory of Gene Regulation
and Signal Transduction
Department of Pharmacology
School of Medicine
University of California, San Diego
La Jolla, California, U.S.A.
Email: e1gallagher@ucsd.edu
Chapter 1

J. Silvio Gutkind
Oral and Pharyngeal Cancer Branch
National Institute of Dental
and Craniofacial Research
National Institutes of Health
Bethesda, Maryland, U.S.A.
Email: sg39v@nih.gov
Chapter 2

Steven X. Hou
The Laboratory of Immunobiology
National Cancer Institute at Frederick
Frederick, Maryland, U.S.A.
Email: shou@mail.ncifcrf.gov
Chapter 5

Michael Karin
Laboratory of Gene Regulation
and Signal Transduction
Department of Pharmacology
School of Medicine
University of California, San Diego
La Jolla, California, U.S.A.
Email: karinoffice@ucsd.edu
Chapter 1

Toshiaki Katada
Department of Physiological Chemistry
Graduate School
of Pharmaceutical Sciences
University of Tokyo
Bunkyo-ku, Tokyo, Japan
Email: katada@mol.f.u-tokyo.ac.jp
Chapter 4

Zheng-gang Liu
Cell and Cancer Biology Branch
Center for Cancer Research
National Cancer Institute
National Institutes of Health
Bethesda, Maryland, U.S.A.
Email: zgliu@helix.nih.gov
Chapter 7

Changwan Lu
The Laboratory of Immunobiology
National Cancer Institute at Frederick
Frederick, Maryland, U.S.A.
Email: cwlu@mail.ncifcrf.gov
Chapter 5

Maria Julia Marinissen
Instituto de Investigaciones Biomedicas
A. Sols UAM-CSIC
Departamento de Bioquimica
Facultad de Medicina
Universidad Autonoma de Madrid
Madrid, Spain
Email: mjmarinissen@iib.uam.es
Chapter 2

Hiroshi Nishina
Department of Developmental
and Regenerative Biology
Medical Research Institute
Tokyo Medical and Dental University
Bunkyo-ku, Tokyo, Japan
Email: nishina.dbio@mri.tmd.ac.jp
Chapter 4

Kanaga Sabapathy
Laboratory of Molecular Carcinogenesis
Division of Cellular
and Molecular Research
National Cancer Centre
and
Department of Biochemistry
National University of Singapore
Singapore
Email: cmrksb@nccs.com.sg
Chapter 3

Yoshitaka Satoh
Medical Chemistry
Celgene
San Diego, California, U.S.A.
Email: ysatoh@celgene.com
Chapter 8

Han-Ming Shen
Department of Community
Occupational and Family Medicine
Faculty of Medicine
National University of Singapore
Singapore
Email: cofshm@nus.edu.sg
Chapter 7

PREFACE

Signal transduction is one of the most exciting research areas in modern biology as it deals with how information flows from the extracellular environment into a living cell to change its metabolism, genotype and phenotype. With the completion of the genomes of human and several other species, it becomes even more important to elucidate the molecular mechanisms that govern cellular functions. The intracellular signaling network, which is composed of many signaling pathways, regulates most, if not all, fundamental cellular activities, from proliferation to programmed cell death (apoptosis). Over the past decade, many signaling pathways have been "mapped' out. Nowadays, we know in great detail about how a specific signal is transmitted via specific signaling pathways, from the membrane to the nucleus, to change the functions of a cell. In this book, a group of experts present a comprehensive review of one of such signaling pathways, the JNK signaling pathway.

In the first chapter, Ewen Gallagher and Michael Karin discuss the discovery of JNK. About a decade ago, JNK was born in Michael Karin's laboratory. Subsequently, JNKs were molecularly cloned by the laboratories of Michael Karin and Roger J. Davis. Simultaneously, the laboratory of Joseph A. Avruch identified and cloned the rat homologues of JNK, the SAPKs (stress-activated protein kinases). In addition, these authors present a comprehensive review of the biological functions of JNK. In the next chapter, Maria Julia Marinissen and J. Silvio Gutkind present an extensive review of the regulation of JNK by small GTP-binding proteins of the Rho family. These authors discussed how JNK is regulated by Rac, Cdc42 and Rho proteins in a cell type- and stimulus-specific manner.

The next three chapters focus on the genetic analysis of the biological functions of JNK and its upstream kinase JNKK (MKK4/MKK7). Using knockout mouse models, Kanaga Sabapathy presents an extensive discussion about the physiological roles of different JNKs, and Hiroshi Nishina and Toshiaki Katada present a comprehensive review of the biological functions of JNKs and its upstream kinase JNKKs (MKK4/MKK7). Using Drosophila as a model system, Changwan Lu and Steven X. Hou present an extensive review of the versatile roles of JNK in Drosophila, from embryonic dorsal closure, larval thorax closure, adult wound healing, planar cell polarity, immune response, synaptic plasticity, neuronal cargo transport, to apoptosis and lifespan.

The role of JNK in cell death is an intensively studied and also highly controversial topic in JNK biology. Anning Lin presents a comprehensive review of the role of JNK in both apoptosis and cell survival. The author discusses the molecular mechanisms underlying the pro- and anti-apoptotic role of JNK and proposes the "breaking the brake on apoptosis" model. Han-Ming Shen and Zheng-gang Liu present an extensive review of the regulation of JNK by reactive oxygen species (ROS) and the role of JNK in ROS-mediated cell death. The authors discuss the signaling mechanisms of ROS-mediated JNK activation and the role of JNK in ROS-mediated cell death (both apoptosis and necrosis).

In the last chapter, Brydon Bennett and Yoshitaka Satoh present an extensive review of the search for the inhibitors of JNK, based on the belief that deregulation of JNK activity may be involved in human diseases. The authors report that the first JNK inhibitors are actually now entering clinical trials to determine safety and efficacy limits in humans. It is possible that small molecules that can specifically inhibit JNK activity could be potential future therapy for certain human diseases.

I hope that this book serves as an important resource for graduate students, postdocs, and basic and clinical researchers who are interested in understanding the biological functions of the intracellular signaling network in general, and the JNK signaling pathway in particular.

Anning Lin
Chicago, Illinois, U.S.A.

CHAPTER 1

The Discovery of JNK and Its Biological Functions

Ewen Gallagher and Michael Karin*

Summary

The c-Jun amino-terminal kinases (JNKs) are critical regulators of various aspects of mammalian physiology, including: cell proliferation, survival and death. Originally the JNKs were isolated as upstream kinases that regulate c-Jun, a component of the AP-1 transcription factor. JNK-mediated phosphorylation enhances c-Jun transcriptional activity and is a major contributor to the activation of AP-1 by extracellular stimuli, such as UV light or the proinflammatory cytokine tumor necrosis factor-α (TNF-α). After the initial isolation and cloning of JNKs 1 to 3, and the subsequent biochemical description of their function (which has included the crystallization of JNK1 and JNK3), targeted disruption of the *Jnk* loci in mice has proved to be a critical tool for a more complete understanding of their physiological functions. It is now known that the JNKs play important roles in numerous cellular processes, including: $CD4^+$ T cell function in the immune system, negative regulation of insulin signaling, control of fat deposition and the proper development of various organs, such as the eyelids and the brain. Also, the JNKs have become prime targets for drug development. As such, this review provides an overview of both the discovery of JNKs, and the characterization of *Jnk* knockout mice.

The Identification of JNK as a Critical Upstream Activator of Jun/AP-1

Transcription factor AP-1 is a DNA binding hetero-dimer that comprises Jun (c-Jun, JunB and JunD), Fos, Maf and ATF family members.[1] These transcription factors belong to the bZIP family and, as such, their dimerization is mediated via their leucine zipper, whereas the recognition of the AP-1 site is mediated through the basic region.[2] In addition to the bZIP domain, many, but not all, of the AP-1 proteins contain transcriptional activation domains, whose activity can be regulated in response to various extracellular stimuli through their phosphorylation.[3] AP-1 transcription factors control the expression of numerous target genes by binding to their promoters and regulating events that lead to transcriptional initiation by eukaryotic RNA polymerase II. One of the most potent stimuli that induces AP-1-dependent gene expression is short wavelength UV light,[4] and it became apparent through phospho-peptide mapping and in vivo metabolic labeling experiments that UV irradiation of cultured cells results in a rapid and substantial increase in c-Jun phosphorylation.[4] This phosphorylation event was suggested to stimulate c-Jun transcriptional activity. Indeed, deletion and site directed mutagenesis experiments identified serines 63 and 73 as the UV responsive sites, that regulate

*Corresponding Author: Michael Karin—Laboratory of Gene Regulation and Signal Transduction, Department of Pharmacology, School of Medicine, University of California, San Diego, 9500 Gilman Drive, La Jolla, California, U.S.A. Email: karinoffice@ucsd.edu

The JNK Signaling Pathway, edited by Anning Lin. ©2006 Eurekah.com.

c-Jun transcriptional activity.[5-8] In addition to UV irradiation, phosphorylation of serines 63 and 73 is stimulated by: growth factors, activated Ras and Src oncoproteins, and occupancy of the T Cell Receptor (TCR).[6,9] Similar phosphorylation sites have been identified in the related molecule JunB, but, unlike c-Jun, JunB is not as responsive to stimuli such as UV radiation.[10]

The next step in understanding c-Jun regulation was the identification of a UV-, TNF-α-, and oncoprotein- responsive protein kinase that phosphorylates it on serines 63 and 73. This was accomplished through the development of a solid-state kinase assay.[11] To develop this novel assay the transcriptional activation domain of c-Jun, encompassing serines 63 and 73, was fused to glutathione S-transferase (GST), expressed in *E. coli*, and the resultant fusion protein was incubated with extracts of nonirradiated, or irradiated, cells, and after washing, γ-^{32}P-ATP was added to test whether c-Jun becomes phosphorylated under these conditions. The results were striking: UV irradiated cells were found to contain a potent kinase activity which was bound to the c-Jun activation domain and phosphorylated it on serines 63 and 73.[11] Subsequently, this assay was used to demonstrate activation of JNKs in T cells.[9] Biochemical characterization, using in-gel kinase assays, led to the identification of two JNK isoforms (46 and 56 kDa in size), known as JNK1 and JNK2, and these have been shown to be distinct from ERK1 and ERK2, which, at the time, were the only known members of the Mitogen-Activated Protein Kinase (MAPK) group in mammals.[11] Yet, based on their activation by extracellular stimuli, and the ability to phosphorylate serines or threonines followed by a proline, which is strictly required, the JNKs were predicted to be new members of the MAPK group of signal transducing protein kinases, a prediction that was confirmed by the molecular cloning of the cDNAs for JNK1 and JNK2.[12,13] The activity of the recombinant proteins encoded by these cDNAs was shown to be stimulated by the same agonists found to activate the endogenous JNK activity detected by the solid state kinase assay. Furthermore, the substrate specificity of the recombinant and endogenous proteins was shown to be identical, with a clear preference for phosphorylation of c-Jun at serines 63 and 73.[12,13] Subsequent work led to the identification of JNK3, whose expression is restricted to neuronal, and cardiac cells.[14] These same kinases were identified by Kyriakis and coworkers, who named them Stress Activated Protein Kinases (SAPKs).[15,16]

MAPKs are conserved throughout eukaryotic evolution, from yeast to humans, and are commonly activated via a three-tier phosphorylation cascade.[3] Thus, MAPKs are activated by MAPK kinases (MKKs, JNKKs or MAP2Ks), and the latter are activated by MKK kinases (MEKKs, or MAP3Ks). MKK4 preferentially phosphorylates tyrosine residue 185 in the JNK1 activation motif (TPY), whereas MKK7 preferentially phosphorylates threonine residue 183.[17,18] The dual specificity MAP2Ks of the JNK pathway, MKK4 and MKK7, are activated by their upstream MAP3Ks through phosphorylation on serine or threonine residues in the activation loop (MKK4 is phosphorylated on residues serine 257 and threonine 265, and MKK7 is phosphorylated on residues serine 271 and threonine 275). A much larger number of MAP3Ks have been identified that can activate JNK, which are likely to act in a cell type and stimulus specific manner.[19] The first and most potent JNK-activating MAP3K to be identified was MEKK1.[20] Since then, however, many more MAP3Ks, which can activate JNK, have been identified, including: MEKK2 and MEKK3,[21] MEKK4,[22] Mixed Lineage kinases 2 and 3 (MLK2 and MLK3),[23,24] Dual leucine zipper-bearing kinase (DLK),[25] Tumor progression locus-2 (Tpl-2),[26] TGF-β-activated kinase (TAK1),[27] Apoptosis Signal Regulating Kinases 1 and 2 (ASK1 and ASK2),[28,29] and Thousand and one amino acid kinases 1 and 2 (Tao1 and Tao2).[30,31]

The Regulation of JNK MAPK Cascades

The JNKs, like other MAPKs, are organized into signaling modules, and this allows their highly coordinated and specific activation in response to stimuli.[32] While there are three MAPKs in the JNK pathway and two MAP2Ks, there are many MAP3Ks (as listed above). The presence of a large number of JNK-activating MAP3Ks in mammals may give the benefit of

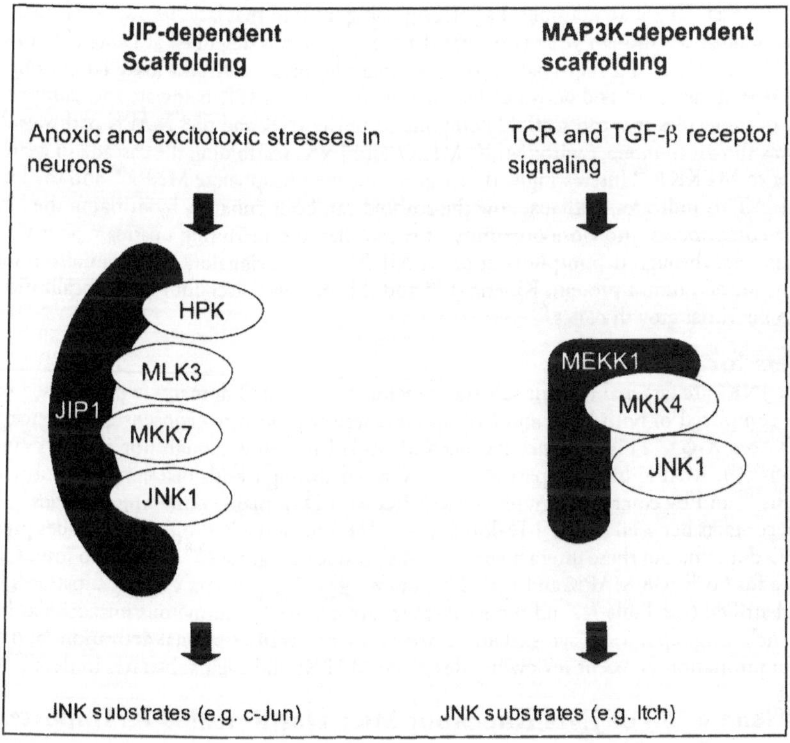

Figure 1. Both JIP1 and MEKK1 can scaffold elements of the JNK MAPK pathway.

improved signal specificity to diverse types of stress stimuli. Further diversity and complexity is introduced by the recent finding that JNK1 is the major isoform responsible for c-Jun amino-terminal phosphorylation, whereas JNK2 may be a negative regulator of this activity.[33] JNK signaling modules are formed by scaffolding proteins, and these usually comprise the JNK Interacting Proteins- (JIP) like scaffolds,[34] and the MKKs, that have an amino-terminal extension that allows them to interact with the downstream MAPK and the upstream MAP3K (this is illustrated in Fig. 1).[35,36] In addition, signaling specificity is found at the level of MAP3Ks interactions with MKKs. MEKK1 preferentially binds to MKK4, whereas MEKK2, a related MAP3K, favors binding to MKK7.[35,37] Molecular scaffolding, most likely, confers improved signaling specificity and integration of JNK signaling in vivo, though it may also limit signal amplification.[38-40] Several different JNK scaffolding molecules have now been identified, including: β-arrestin-2,[41] CRK2,[42] Filamin,[43] IKAP,[44] JIP1,[45] JIP2,[46] JIP3,[47,48] and JIP4.[49] The function of these molecules has been extensively described in a recent review.[36]

MEKK1 is an example of a MAP3K that can scaffold elements of its own signaling cascade, via its kinase domain and long amino-terminal domain.[36] It can form complexes with MKK4,[35] and JNK1 or JNK2.[50] The phosphorylation state of MEKK1 may influence its ability to interact with components of the JNK MAPK pathway, as JNKs are recruited to the MEKK1 signaling module following its activation by stress stimuli.[51] In addition, MEKK1 shows a striking localization to focal adhesions and the actin cytokskeleton, and can interact with the cytoskeletal factor α-actinin,[52] Rho family GTPases,[53,54] and p115 RhoGAP.[55] In addition, it contains a Plant Homology Domain (PHD) that has in vitro E3 ubiquitin ligase activity, and can mediate the ubiquitination and proteolysis of itself,[56] and of substrates such as ERK2.[57]

By contrast, JIP1 is an example of a JNK signaling scaffold that lacks its own kinase activity. Initially identified through yeast two-hybrid screening, it was described as being able to suppress JNK activity.[58] Further analysis of this molecule, under different assay conditions, revealed that it can bind and activate elements of the JNK MAPK pathway, and comprises a family of molecules that can scaffold components of the JNK and p38 MAPK pathways.[45,59] JIP1 was shown to interact with: MLK, MKK7 and JNK, scaffolding the cascade in a similar manner to MEKK1.[36] Interestingly, JIP1 can recruit the phosphatase MKP7[60] and the kinase AKT-1.[61] This indicates, perhaps, how the scaffold can both enhance, by bringing the kinase cascade components into close proximity of each other and recruiting upstream activators,[61] and suppress, through dephosphorylation by MKP7,[60] JNK signaling. JIP1 can also interact with the kinesin motor protein, Kinesin-1,[62] and this may well account for the localization of JIP1 to neuronal growth cones.[63]

Targets for JNKs

The JNKs are targeted to their substrates by motifs termed D domains, a peptide sequence that is comprised of both basic and hydrophobic residues, with the consensus sequence: $(K/R)_{2-3}$-X_{1-6}-φ-X-φ.[64] This sequence has been identified in many, though not all (for example PEA-15[65,66]), MAPK binding partners, and verified through both deletion and point mutagenesis,[10] and by co-crystallography of MAPKs with D domain containing peptides.[67] The critical contacts between the JIP1 D domain and JNK1 lie in the hydrophobic residues present in the D domain, and these interact with the MAPK docking groove.[68] This is also found to be the case for both p38 MAPK and ERK2.[67] An ever growing plethora of JNK substrates have been identified (see Table 1), and these substrates are involved in numerous intracellular functions, including: apoptosis, cytoskeleton rearrangements, transcriptional activation, and protein ubiquitination (a recent review has described MAPKs and their substrates in detail[69]).

The Generation of JNK Knockout Mice Has Revealed Its Importance in Mammalian Biology

JNKs in the Immune System and the Regulation of T Lymphocyte Differentiation

Simultaneous cross-linking of the TCR, with antibodies to its CD3 subunit and the CD28 costimulatory receptor, results in a rapid and robust activation of JNKs and c-Jun phosphorylation, on serines 63 and 73.[9] Engagement of either receptor alone has only a minor effect on JNK activity. It was, therefore, not a surprise when JNK knockout mice were found to have defects in T cell functions. Although thymic development and positive selection of T cells within the thymus are largely unaltered in *Jnk1*$^{-/-}$ or *Jnk2*$^{-/-}$ mice, the JNKs may play a role in the depletion of thymocytes and in activation induced cell death.[70,71] The most striking T cell phenotypes exhibited by JNK-deficient mice relate to the extra-thymic differentiation of $CD4^+$ T cells.[72]

$CD4^+$ T helper (Th) cells undergo differentiation into two distinct classes of effector T cells, depending on their profiles of secreted cytokines. Th1 cells are defined as those that secrete the cytokines: Interferon-γ (IFN-γ), and TNF-β. The main role of these T cells is in controlling immune system responses to viral infections and intracellular pathogens, whose antigens are presented via class II MHC molecules.[73] Th2 cells, by contrast, are characterized by the secretion of Interleukins: 4, 5, 9, 10, 11 and 13,[72] and are involved in orchestrating humoral immune responses to extracellular pathogens, whose antigens are presented via class I MHC molecules.[73] $CD4^+$ T cells from *Jnk1*$^{-/-}$ mice were found to be polarized towards a Th2 profile of secreted cytokines, after undergoing in vitro differentiation.[74] By contrast, *Jnk2*$^{-/-}$ mice have a T lymphocyte phenotype of a polarized Th1 response, but a normal Th2 response.[75] Perhaps surprisingly, *Jnk1*$^{-/-}$/*Jnk2*$^{-/-}$ double knockout T cells, generated by in vitro differentiation of double mutant embryonic stem (ES) cells, show polarization towards Th2 cytokine

Table 1. JNK substrates and their cellular functions

JNK Substrate	Cellular Function	Reference
IRS-1	Insulin signaling	91
Bcl-2	Apoptosis	106
Bcl-XL	Apoptosis	107
Bim	Apoptosis	108
Bad	Apoptosis	109
DENN/MADD	Death domain/ neuronal signaling	110
NFATc1	Transcription factor	111
ATF2	Transcription factor	112
c-Jun	Transcription factor	11
STAT3	Transcription factor	113
p53	Transcription factor	114
Elk-1	Transcription factor	115
Net	Transcription factor	116
p90RSK	Transcription factor	117
JunB	Transcription factor	10
p66ShcA	Adaptor	118
LAT	Adaptor	119
Sab	Mitochondrial protein	120
Paxillin	Cytoskeleton	104
MAP1/ MAP2	Cytoskeleton	100
TAU	Cytoskeleton	121
Itch	E3 ubiquitin ligase	87
Heat shock factor-1	Heat shock	122

secretion, similar to *Jnk1*$^{-/-}$T cells.[76] Similarly, *Jnk2*$^{-/-}$ T cells that express dominant negative JNK1 display the same Th2 polarization as *Jnk1*$^{-/-}$ single knockout CD4$^+$ T cells.[76] Neither of the T cells derived from these two sources is defective in IL-2 cytokine production, a cytokine important in T cell proliferation. Thus, a clear role has been shown for JNKs in controlling the balance of CD4$^+$ Th cell differentiation, though they do not appear to play a significant role in naïve Th cell activation.[77]

More recently, defects in CD8$^+$ T cells from *Jnk* deficient mice were found following immunization with viral particles.[78,79] As with CD4$^+$ T cells, the phenotypes of CD8$^+$ T cells derived from either *Jnk1*$^{-/-}$ or *Jnk2*$^{-/-}$ mice are divergent. CD8$^+$ T cells from *Jnk1*$^{-/-}$ mice were found to proliferate to a lesser extent than those from Wild-type (WT) mice following viral challenge.[78,79] By contrast, *Jnk2*$^{-/-}$ mice show increased numbers of CD8$^+$ T cells following viral infection.[78,79] Why CD8$^+$ T cells behave very differently from CD4$^+$ T cells in respect to the functions of JNK1 and JNK2 is currently unknown.

One answer to this puzzle may lie in the effects on Jun family members, the transcription factor substrates of JNKs, and some of these are key regulators of T cell differentiation in vivo. The JunB protein was found to be important for the production of Th2 effector cells. Its absence prevents Th2, but not Th1 differentiation,[80] and its overexpression pushes naïve CD4$^+$ T cells towards a phenotype without a marked effect on Th1 differentiation.[81] The genetic analysis of the JNK activating kinases, MKK4 and MKK7, is complicated by the embryonic lethality associated with their absence.[82] However, radiation chimeras reconstituted with *MKK4*$^{-/-}$ hematopoietic cells exhibit abnormalities in their peripheral T cell population,[83] and deficient TCR signaling and apoptosis.[84] A similar analysis revealed that the loss of *MKK7* results in thymocyte hyperproliferation.[85]

MEKK1 kinase-deficient mice (*Mekk1*$^{\Delta KD}$)[86] also display aberrant Th cell differentiation of CD4$^+$ T cells, and exhibit much lower levels of JNK activity after engagement with anti-CD3 and anti-CD28 compared to WT mice. Correspondingly, these MEKK1-deficient T cells show a very strong Th2 bias during in vitro differentiation,[87] similar to the one exhibited by T cells from *Jnk1*$^{-/-}$ mice,[88] or JunB overexpressing CD4$^+$ T cells.[81] Interestingly, *Mekk1*$^{\Delta KD/\Delta KD}$ CD4$^+$ T cells have elevated levels of JunB expression, suggesting that the JNK cascade is involved in the control of JunB abundance through the regulation of Itch activity.[87]

JNK1 Controls Obesity and Insulin Signaling

Outside of the immune system, the JNKs also play an important role in metabolic control.[89] Consumption of a high fat diet (HFD) or genetic obesity interfere with proper insulin signaling and lead to a condition termed insulin resistance, that if left untreated, can result in type II diabetes.[89] It was suggested that the down-modulation of insulin signaling is due to the production of low levels of inflammatory mediators, which activate a protein kinase cascade that leads to phosphorylation of Insulin Receptor Substrate (IRS1), an essential adaptor for Insulin Receptor (IR) signaling.[89] Normally, the binding of insulin to IR results in IRS1 phosphorylation on tyrosine residues, a modification that is required for recruitment and activation of various effectors.[90] However, in response to obesity or inflammation, IRS1 becomes phosphorylated at serine 307, a modification that interferes with its recruitment to the IR and subsequent phosphorylation on tyrosine residues.[91] It was demonstrated that, in vitro, JNKs can phosphorylate IRS1 on serine 307, and that this correlates with reduced IR-mediated IRS1 tyrosine phosphorylation.[92] The known ability of the JNKs to undergo activation in response to proinfamatory cytokines, such as TNF-α or IL-1,[13] suggested that they may be involved in phosphorylation of IRS1 serine 307, a proline-directed phosphorlyation site. Indeed, *Jnk1*$^{-/-}$ mice were found to be much more responsive than WT mice when placed on a HFD, or even after crossing to genetically obese mice.[93] In addition, *Jnk1*$^{-/-}$ mice show reduced fat build up, suggesting an additional role for JNKs in the control of fat disposition or adipocyte differentiation. *Jnk2*$^{-/-}$ mice, by contrast, appear indistinguishable from the WT.

Importantly, tissues isolated from *Jnk1*$^{-/-}$ mice placed on a HFD show much lower levels of IRS1 serine 307 phosphorylation than tissues from similarly treated WT mice.[93] It is also of interest that mutations in the *Jip1* gene have been found in patients with a genetic form of type II diabetes.[94] This suggests that a JIP1 organized JNK1 cascade plays an important role in the regulation of insulin signaling.

JNKs and Neuronal Functions

JNKs have a well-documented function in the mammalian nervous system. In mouse embryos JNK1 and JNK2 are expressed in neuronal tissues after embryonic day (E) 7, and JNK3 is seen from around E11.[95] Expression is found to be high in both the fore and hind-brain regions of mice. Despite the high, and restricted, expression of JNK3 in the neuronal tissues in comparison to the other JNKs, *Jnk3*$^{-/-}$ mice have a brain that appears morphologically normal.[96] However, these mice, but not *Jnk1*$^{-/-}$ or *Jnk2*$^{-/-}$ single knockouts, are resistant to treatment with excitotoxic agents, such as kainic acid, which cause seizures and increased apoptosis of hypocampal neurons. These observations may prove relevant to Parkinson's disease and other neurodegenerative disorders in humans, because *Jnk3*$^{-/-}$ mice treated with the neurotoxin 1-methyl-4-phenyl-1,2,4,6 tetrahydropyridine (MPTP), that replicates many of clinical symptoms seen in Parkinson's syndrome, including loss of tyrosine hydroxylase positive nigral cells, show much lower levels of neuronal cell death.[97] Surprisingly, this resistance is also observed in *Jnk2*$^{-/-}$ mice. Indeed, the *Jnk2*$^{-/-}$/*Jnk3*$^{-/-}$ double knockout animals exhibit an even greater level of resistance to MPTP than either of the single knockouts.[97] Also, the JNKs have been demonstrated to be involved in cerebral ischemia-hypoxia brought about by carotid artery ligation and hypoxia.[98] In this animal model elevated JNK activation, and caspase cleavage, were observed. When subjected to this experimental model *Jnk3*$^{-/-}$ mice were found to display resistance to cerebral ischemia–hypoxia, in comparison to WT, *Jnk1*$^{-/-}$, or *Jnk2*$^{-/-}$ mice.[99]

Although the *Jnk1*$^{-/-}$ or *Jnk2*$^{-/-}$ mouse brain initially appears normal, a detailed study examining the development of the nervous systems in these mice revealed that the axons forming the anterior commissure are absent in *Jnk1*$^{-/-}$, but not in *Jnk2*$^{-/-}$ mice, after 2 to 3 months of age.[100] These findings led to the discovery of progressive neurite degeneration in *Jnk1*$^{-/-}$ mice, which is associated with a shortening of neuronal microtubules.[100] This defect is due to a deficiency in phosphorylation of the MAP1 and MAP2 microtubule associated proteins that maintain microtubule stability. These proteins are phosphorylated more efficiently by JNK1 than JNK2, and JNK1-dependent phosphorylation enhances their interaction with microtubules.[100] Another JNK-dependent neuronal function was revealed by *Jnk1*$^{-/-}$/*Jnk2*$^{-/-}$ double knockout mice, which die at around E11, and these develop, during embryogenesis, hind-brain exencepthaly due to defective neural tube closure.[95] Increased caspase activation, and apoptosis, are seen in critical parts of the forebrain from around E10.5 of embryogenesis, though this cell death is not observed in the mid- or hind- brain of these mice. The neural tube closure defect found in the *Jnk1*$^{-/-}$/*Jnk2*$^{-/-}$ double knockout embryos may then be explained, in part, by the reduced levels of apoptosis found in regions of the hindbrain folds.[95] Thus, JNKs may critically play both pro- and anti-apoptotic roles in the developing mammalian nervous system.

JNKs Are Required for Mammalian Eyelid Fusion, Keratinocyte Migration, and Wound Healing

During mammalian embryogenesis the eyelids fuse in a process reminiscent of dorsal closure seen *Drosophila melanogaster*, which is known to be JNK-dependent.[101,102] As with *Drosophila*, a critical role has been found for JNKs in mammalian eyelid fusion during embryonic development. In mice eyelid fusion begins at E15.5 to E16.5. Interestingly, MEKK1 is highly expressed at the growing edge of the eyelid and loss of its catalytic activity disrupts the fusion process causing the *Mekk1*$^{\Delta KD/\Delta KD}$ mice to be born with an Eye Open at Birth (EOB) phenotype.[86] Mechanistically the eyelid fusion process, like dorsal closure in *Drosophila*, involves the fusion of two migrating epithelial sheets.[103] This migratory process is associated with the formation of F-actin cables, and is found to be defective in the eyelids of *Mekk1*$^{\Delta KD/\Delta KD}$ mice.[86] A similar defect in migration and F-actin polymerization is seen in cultured keratinocytes in which these processes were induced by incubation with TGF-β (or its family member activin). Interestingly, *Mekk1*$^{\Delta KD/\Delta KD}$ keratinocytes migrate normally in response to TGF-α, which activates the EGF receptor.[86] These results indicate that MEKK1 is required for the transmission of signals generated at TGF-β receptors. These signals are transmitted to JNK, whose activation is required for F-actin polymerization. This can be inferred from the defective JNK activation seen in keratinocytes and eyelids of either *Mekk1*$^{\Delta KD/\Delta KD}$ mice or *Mekk1*$^{+/\Delta KD}$/*Jnk1*$^{+/-}$ double heterozygote mice. This phenotype is not found in single heterozygote mice, which have normal activation of JNKs.[86] TGF-β-induced F-actin polymerization is also blocked by the low molecular weight JNK inhibitor. Biochemically, the mechanism by which JNK stimulates F-actin polymerization may lie in the phosphorylation of paxillin.[104] JNKs phosphorylate paxillin, a focal adhesion associated protein, at serine 178, and mutation of this residue inhibits cell migration when introduced into cell lines.

Perspectives

Ten years after their initial discovery and cloning, our knowledge of JNKs role in the mammalian physiology has greatly increased through both genetic and biochemical investigations. The JNKs have a large number of intracellular targets, and, by phosphorylation of their substrates, regulate numerous cellular processes. Knockout mice have proved to be one of the most important tools in elucidating the physiological role of JNKs. Investigations using *Jnk* deficient mice, have, so far, identified roles for this family of MAPKs in: neuronal tissues, eyelid fusion during embryogenesis, obesity and insulin resistance, and T cell differentiation. *Jnk* knockout mice will, most likely, continue to be key in the identification of additional roles for JNKs in the mammalian biology.

References

1. Shaulian E, Karin M. AP-1 in cell proliferation and survival. Oncogene 2001; 20(19):2390-2400.
2. Angel P, Karin M. The role of Jun, Fos and the AP-1 complex in cell-proliferation and transformation. Biochim Biophys Acta 1991; 1072(2-3):129-157.
3. Karin M. The regulation of AP-1 activity by mitogen-activated protein kinases. Philos Trans R Soc Lond B Biol Sci 1996; 351(1336):127-134.
4. Devary Y, Gottlieb RA, Smeal T et al. The mammalian ultraviolet response is triggered by activation of Src tyrosine kinases. Cell 1992; 71(7):1081-1091.
5. Pulverer BJ, Kyriakis JM, Avruch J et al. Phosphorylation of c-jun mediated by MAP kinases. Nature 1991; 353(6345):670-674.
6. Smeal T, Binetruy B, Mercola DA et al. Oncogenic and transcriptional cooperation with Ha-Ras requires phosphorylation of c-Jun on serines 63 and 73. Nature 1991; 354(6353):494-496.
7. Binetruy B, Smeal T, Karin M. Ha-Ras augments c-Jun activity and stimulates phosphorylation of its activation domain. Nature 1991; 351(6322):122-127.
8. Smeal T, Binetruy B, Mercola D et al. Oncoprotein-mediated signalling cascade stimulates c-Jun activity by phosphorylation of serines 63 and 73. Mol Cell Biol 1992; 12(8):3507-3513.
9. Su B, Jacinto E, Hibi M et al. JNK is involved in signal integration during costimulation of T lymphocytes. Cell 1994; 77(5):727-736.
10. Kallunki T, Deng T, Hibi M et al. c-Jun can recruit JNK to phosphorylate dimerization partners via specific docking interactions. Cell 1996; 87(5):929-939.
11. Hibi M, Lin A, Smeal T et al. Identification of an oncoprotein- and UV-responsive protein kinase that binds and potentiates the c-Jun activation domain. Genes Dev 1993; 7(11):2135-2148.
12. Derijard B, Hibi M, Wu IH et al. JNK1: A protein kinase stimulated by UV light and Ha-Ras that binds and phosphorylates the c-Jun activation domain. Cell 1994; 76(6):1025-1037.
13. Kallunki T, Su B, Tsigelny I et al. JNK2 contains a specificity-determining region responsible for efficient c-Jun binding and phosphorylation. Genes Dev 1994; 8(24):2996-3007.
14. Gupta S, Barrett T, Whitmarsh AJ et al. Selective interaction of JNK protein kinase isoforms with transcription factors. Embo J 1996; 15(11):2760-2770.
15. Kyriakis JM, Banerjee P, Nikolakaki E et al. The stress-activated protein kinase subfamily of c-Jun kinases. Nature 1994; 369(6476):156-160.
16. Kyriakis JM, Woodgett JR, Avruch J. The stress-activated protein kinases. A novel ERK subfamily responsive to cellular stress and inflammatory cytokines. Ann NY Acad Sci 1995; 766:303-319.
17. Lawler S, Fleming Y, Goedert M et al. Synergistic activation of SAPK1/JNK1 by two MAP kinase kinases in vitro. Curr Biol 1998; 8(25):1387-1390.
18. Lisnock J, Griffin P, Calaycay J et al. Activation of JNK3 alpha 1 requires both MKK4 and MKK7: Kinetic characterization of in vitro phosphorylated JNK3 alpha 1. Biochemistry 2000; 39(11):3141-3148.
19. Davis RJ. Transcriptional regulation by MAP kinases. Mol Reprod Dev 1995; 42(4):459-467.
20. Minden A, Lin A, McMahon M et al. Differential activation of ERK and JNK mitogen-activated protein kinases by Raf-1 and MEKK. Science 1994; 266(5191):1719-1723.
21. Blank JL, Gerwins P, Elliott EM et al. Molecular cloning of mitogen-activated protein/ERK kinase kinases (MEKK) 2 and 3. Regulation of sequential phosphorylation pathways involving mitogen-activated protein kinase and c-Jun kinase. J Biol Chem 1996; 271(10):5361-5368.
22. Gerwins P, Blank JL, Johnson GL. Cloning of a novel mitogen-activated protein kinase kinase kinase, MEKK4, that selectively regulates the c-Jun amino terminal kinase pathway. J Biol Chem 1997; 272(13):8288-8295.
23. Tibbles LA, Ing YL, Kiefer F et al. MLK-3 activates the SAPK/JNK and p38/RK pathways via SEK1 and MKK3/6. Embo J 1996; 15(24):7026-7035.
24. Hirai S, Izawa M, Osada S et al. Activation of the JNK pathway by distantly related protein kinases, MEKK and MUK. Oncogene 1996; 12(3):641-650.
25. Fan G, Merritt SE, Kortenjann M et al. Dual leucine zipper-bearing kinase (DLK) activates p46SAPK and p38mapk but not ERK2. J Biol Chem 1996; 271(40):24788-24793.
26. Salmeron A, Ahmad TB, Carlile GW et al. Activation of MEK-1 and SEK-1 by Tpl-2 proto-oncoprotein, a novel MAP kinase kinase kinase. Embo J 1996; 15(4):817-826.
27. Yamaguchi K, Shirakabe K, Shibuya H et al. Identification of a member of the MAPKKK family as a potential mediator of TGF-beta signal transduction. Science 1995; 270(5244):2008-2011.
28. Ichijo H, Nishida E, Irie K et al. Induction of apoptosis by ASK1, a mammalian MAPKKK that activates SAPK/JNK and p38 signaling pathways. Science 1997; 275(5296):90-94.
29. Wang XS, Diener K, Tan TH et al. MAPKKK6, a novel mitogen-activated protein kinase kinase kinase, that associates with MAPKKK5. Biochem Biophys Res Commun 1998; 253(1):33-37.

30. Chen Z, Hutchison M, Cobb MH. Isolation of the protein kinase TAO2 and identification of its mitogen-activated protein kinase/extracellular signal-regulated kinase kinase binding domain. J Biol Chem 1999; 274(40):28803-28807.
31. Hutchison M, Berman KS, Cobb MH. Isolation of TAO1, a protein kinase that activates MEKs in stress-activated protein kinase cascades. J Biol Chem 1998; 273(44):28625-28632.
32. Davis RJ. Signal transduction by the JNK group of MAP kinases. Cell 2000; 103(2):239-252.
33. Sabapathy K, Hochedlinger K, Nam SY et al. Distinct roles for JNK1 and JNK2 in regulating JNK activity and c-Jun-dependent cell proliferation. Molecular Cell 2004.
34. Davis RJ. Signal transduction by the c-Jun N-terminal kinase. Biochem Soc Symp 1999; 64:1-12.
35. Xia Y, Wu Z, Su B et al. JNKK1 organizes a MAP kinase module through specific and sequential interactions with upstream and downstream components mediated by its amino-terminal extension. Genes Dev 1998; 12(21):3369-3381.
36. Morrison DK, Davis RJ. Regulation of MAP kinase signaling modules by scaffold proteins in mammals. Annu Rev Cell Dev Biol 2003; 19:91-118.
37. Cheng J, Yang J, Xia Y et al. Synergistic interaction of MEK kinase 2, c-Jun N-terminal kinase (JNK) kinase 2, and JNK1 results in efficient and specific JNK1 activation. Mol Cell Biol 2000; 20(7):2334-2342.
38. Bagowski CP, Besser J, Frey CR et al. The JNK cascade as a biochemical switch in mammalian cells: Ultrasensitive and all-or-none responses. Curr Biol 2003; 13(4):315-320.
39. Bagowski CP, Ferrell Jr JE. Bistability in the JNK cascade. Curr Biol 2001; 11(15):1176-1182.
40. Ferrell Jr JE. Self-perpetuating states in signal transduction: Positive feedback, double-negative feedback and bistability. Curr Opin Cell Biol 2002; 14(2):140-148.
41. McDonald PH, Chow CW, Miller WE et al. Beta-arrestin 2: A receptor-regulated MAPK scaffold for the activation of JNK3. Science 2000; 290(5496):1574-1577.
42. Girardin SE, Yaniv M. A direct interaction between JNK1 and CrkII is critical for Rac1-induced JNK activation. Embo J 2001; 20(13):3437-3446.
43. Marti A, Luo Z, Cunningham C et al. Actin-binding protein-280 binds the stress-activated protein kinase (SAPK) activator SEK-1 and is required for tumor necrosis factor-alpha activation of SAPK in melanoma cells. J Biol Chem 1997; 272(5):2620-2628.
44. Holmberg C, Katz S, Lerdrup M et al. A novel specific role for I kappa B kinase complex-associated protein in cytosolic stress signaling. J Biol Chem 2002; 277(35):31918-31928.
45. Whitmarsh AJ, Cavanagh J, Tournier C et al. A mammalian scaffold complex that selectively mediates MAP kinase activation. Science 1998; 281(5383):1671-1674.
46. Negri S, Oberson A, Steinmann M et al. cDNA cloning and mapping of a novel islet-brain/JNK-interacting protein. Genomics 2000; 64(3):324-330.
47. Ito M, Yoshioka K, Akechi M et al. JSAP1, a novel jun N-terminal protein kinase (JNK)-binding protein that functions as a Scaffold factor in the JNK signaling pathway. Mol Cell Biol 1999; 19(11):7539-7548.
48. Kelkar N, Gupta S, Dickens M et al. Interaction of a mitogen-activated protein kinase signaling module with the neuronal protein JIP3. Mol Cell Biol 2000; 20(3):1030-1043.
49. Lee CM, Onesime D, Reddy CD et al. JLP: A scaffolding protein that tethers JNK/p38MAPK signaling modules and transcription factors. Proc Natl Acad Sci USA 2002; 99(22).14189-14194.
50. Xu S, Cobb MH. MEKK1 binds directly to the c Jun N-terminal kinases/stress-activated protein kinases. J Biol Chem 1997; 272(51):32056-32060.
51. Gallagher ED, Xu S, Moomaw C et al. Binding of JNK/SAPK to MEKK1 is regulated by phosphorylation. J Biol Chem 2002; 277(48):45785-45792.
52. Christerson LB, Vanderbilt CA, Cobb MH. MEKK1 interacts with alpha-actinin and localizes to stress fibers and focal adhesions. Cell Motil Cytoskeleton 1999; 43(3):186-198.
53. Fanger GR, Johnson NL, Johnson GL. MEK kinases are regulated by EGF and selectively interact with Rac/Cdc42. Embo J 1997; 16(16):4961-4972.
54. Gallagher ED, Gutowski S, Sternweis PC et al. RhoA binds to the amino terminus of MEKK1 and regulates its kinase activity. J Biol Chem 2004; 279(3):1872-1877.
55. Christerson LB, Gallagher E, Vanderbilt CA et al. p115 Rho GTPase activating protein interacts with MEKK1. J Cell Physiol 2002; 192(2):200-208.
56. Witowsky JA, Johnson GL. Ubiquitylation of MEKK1 inhibits its phosphorylation of MKK1 and MKK4 and activation of the ERK1/2 and JNK pathways. J Biol Chem 2003; 278(3):1403-1406.
57. Lu Z, Xu S, Joazeiro C et al. The PHD domain of MEKK1 acts as an E3 ubiquitin ligase and mediates ubiquitination and degradation of ERK1/2. Mol Cell 2002; 9(5):945-956.
58. Dickens M, Rogers JS, Cavanagh J et al. A cytoplasmic inhibitor of the JNK signal transduction pathway. Science 1997; 277(5326):693-696.

59. Whitmarsh AJ, Davis RJ. Structural organization of MAP-kinase signaling modules by scaffold proteins in yeast and mammals. Trends Biochem Sci 1998; 23(12):481-485.
60. Willoughby EA, Perkins GR, Collins MK et al. The JNK-interacting protein-1 scaffold protein targets MAPK phosphatase-7 to dephosphorylate JNK. J Biol Chem 2003; 278(12):10731-10736.
61. Kim AH, Yano H, Cho H et al. Akt1 regulates a JNK scaffold during excitotoxic apoptosis. Neuron 2002; 35(4):697-709.
62. Verhey KJ, Meyer D, Deehan R et al. Cargo of kinesin identified as JIP scaffolding proteins and associated signaling molecules. J Cell Biol 2001; 152(5):959-970.
63. Whitmarsh AJ, Kuan CY, Kennedy NJ et al. Requirement of the JIP1 scaffold protein for stress-induced JNK activation. Genes Dev 2001; 15(18):2421-2432.
64. Goldsmith EJ, Cobb MH, Chang CI. Structure of MAPKs. Methods Mol Biol 2004; 250:127-144.
65. Whitehurst AW, Robinson FL, Moore MS et al. The death effector domain protein PEA-15 prevents nuclear entry of ERK2 by inhibiting required interactions. J Biol Chem 2004; 279(13):12840-12847.
66. Hill JM, Vaidyanathan H, Ramos JW et al. Recognition of ERK MAP kinase by PEA-15 reveals a common docking site within the death domain and death effector domain. Embo J 2002; 21(23):6494-6504.
67. Chang CI, Xu BE, Akella R et al. Crystal structures of MAP kinase p38 complexed to the docking sites on its nuclear substrate MEF2A and activator MKK3b. Mol Cell 2002; 9(6):1241-1249.
68. Heo YS, Kim SK, Seo CI et al. Structural basis for the selective inhibition of JNK1 by the scaffolding protein JIP1 and SP600125. Embo J 2004; 23(11):2185-2195.
69. Chen Z, Gibson TB, Robinson F et al. MAP kinases. Chem Rev 2001; 101(8):2449-2476.
70. Sabapathy K, Kallunki T, David JP et al. c-Jun NH2-terminal kinase (JNK)1 and JNK2 have similar and stage-dependent roles in regulating T cell apoptosis and proliferation. J Exp Med 2001; 193(3):317-328.
71. Rincon M, Whitmarsh A, Yang DD et al. The JNK pathway regulates the In vivo deletion of immature CD4(+)CD8(+) thymocytes. J Exp Med 1998; 188(10):1817-1830.
72. Flavell RA, Li B, Dong C et al. Molecular basis of T-cell differentiation. Cold Spring Harb Symp Quant Biol 1999; 64:563-571.
73. Medzhitov R, Janeway Jr CA. How does the immune system distinguish self from nonself? Semin Immunol 2000; 12(3):185-188, (discussion 257-344).
74. Dong C, Yang DD, Wysk M et al. Defective T cell differentiation in the absence of Jnk1. Science 1998; 282(5396):2092-2095.
75. Yang DD, Conze D, Whitmarsh AJ et al. Differentiation of CD4+ T cells to Th1 cells requires MAP kinase JNK2. Immunity 1998; 9(4):575-585.
76. Dong C, Yang DD, Tournier C et al. JNK is required for effector T-cell function but not for T-cell activation. Nature 2000; 405(6782):91-94.
77. Flavell RA, Dong C, Davis RJ. Signaling and cell death in lymphocytes. Inflamm Res 2002; 51(2):80-82.
78. Conze D, Krahl T, Kennedy N et al. c-Jun NH(2)-terminal kinase (JNK)1 and JNK2 have distinct roles in CD8(+) T cell activation. J Exp Med 2002; 195(7):811-823.
79. Arbour N, Naniche D, Homann D et al. c-Jun NH(2)-terminal kinase (JNK)1 and JNK2 signaling pathways have divergent roles in CD8(+) T cell-mediated antiviral immunity. J Exp Med 2002; 195(7):801-810.
80. Hartenstein B, Teurich S, Hess J et al. Th2 cell-specific cytokine expression and allergen-induced airway inflammation depend on JunB. Embo J 2002; 21(23):6321-6329.
81. Li B, Tournier C, Davis RJ et al. Regulation of IL-4 expression by the transcription factor JunB during T helper cell differentiation. Embo J 1999; 18(2):420-432.
82. Yang D, Tournier C, Wysk M et al. Targeted disruption of the MKK4 gene causes embryonic death, inhibition of c-Jun NH2-terminal kinase activation, and defects in AP-1 transcriptional activity. Proc Natl Acad Sci USA 1997; 94(7):3004-3009.
83. Swat W, Fujikawa K, Ganiatsas S et al. SEK1/MKK4 is required for maintenance of a normal peripheral lymphoid compartment but not for lymphocyte development. Immunity 1998; 8(5):625-634.
84. Nishina H, Fischer KD, Radvanyi L et al. Stress-signalling kinase Sek1 protects thymocytes from apoptosis mediated by CD95 and CD3. Nature 1997; 385(6614):350-353.
85. Sasaki T, Wada T, Kishimoto H et al. The stress kinase mitogen-activated protein kinase kinase (MKK)7 is a negative regulator of antigen receptor and growth factor receptor-induced proliferation in hematopoietic cells. J Exp Med 2001; 194(6):757-768.
86. Zhang L, Wang W, Hayashi Y et al. A role for MEK kinase 1 in TGF-beta/activin-induced epithelium movement and embryonic eyelid closure. Embo J 2003; 22(17):4443-4454.

87. Gao M, Labuda T, Xia Y et al. Jun turnover is controlled through JNK-dependent phosphorylation of the E3 ligase Itch. Science 2004; 306(5694):271-5.
88. Chen N, Ma W, Huang C et al. Translocation of protein kinase Cepsilon and protein kinase Cdelta to membrane is required for ultraviolet B-induced activation of mitogen-activated protein kinases and apoptosis. J Biol Chem 1999; 274(22):15389-15394.
89. Hotamisligil GS. Inflammatory pathways and insulin action. Int J Obes Relat Metab Disord 2003; 27(Suppl 3):S53-55.
90. White MF. IRS proteins and the common path to diabetes. Am J Physiol Endocrinol Metab 2002; 283(3):E413-422.
91. Aguirre V, Werner ED, Giraud J et al. Phosphorylation of Ser307 in insulin receptor substrate-1 blocks interactions with the insulin receptor and inhibits insulin action. J Biol Chem 2002; 277(2):1531-1537.
92. Aguirre V, Uchida T, Yenush L et al. The c-Jun NH(2)-terminal kinase promotes insulin resistance during association with insulin receptor substrate-1 and phosphorylation of Ser(307). J Biol Chem 2000; 275(12):9047-9054.
93. Hirosumi J, Tuncman G, Chang L et al. A central role for JNK in obesity and insulin resistance. Nature 2002; 420(6913):333-336.
94. Waeber G, Delplanque J, Bonny C et al. The gene MAPK8IP1, encoding islet-brain-1, is a candidate for type 2 diabetes. Nat Genet 2000; 24(3):291-295.
95. Kuan CY, Yang DD, Samanta Roy DR et al. The Jnk1 and Jnk2 protein kinases are required for regional specific apoptosis during early brain development. Neuron 1999; 22(4):667-676.
96. Yang DD, Kuan CY, Whitmarsh AJ et al. Absence of excitotoxicity-induced apoptosis in the hippocampus of mice lacking the Jnk3 gene. Nature 1997; 389(6653):865-870.
97. Hunot S, Vila M, Teismann P et al. JNK-mediated induction of cyclooxygenase 2 is required for neurodegeneration in a mouse model of Parkinson's disease. Proc Natl Acad Sci USA 2004; 101(2):665-670.
98. Kuan CY, Whitmarsh AJ, Yang DD et al. A critical role of neural-specific JNK3 for ischemic apoptosis. Proc Natl Acad Sci USA 2003; 100(25):15184-15189.
99. Haydar TF, Kuan CY, Flavell RA et al. The role of cell death in regulating the size and shape of the mammalian forebrain. Cereb Cortex 1999; 9(6):621-626.
100. Chang L, Jones Y, Ellisman MH et al. JNK1 is required for maintenance of neuronal microtubules and controls phosphorylation of microtubule-associated proteins. Dev Cell 2003; 4(4):521-533.
101. Sluss HK, Han Z, Barrett T et al. A JNK signal transduction pathway that mediates morphogenesis and an immune response in Drosophila. Genes Dev 1996; 10(21):2745-2758.
102. Sluss HK, Davis RJ. Embryonic morphogenesis signaling pathway mediated by JNK targets the transcription factor JUN and the TGF-beta homologue decapentaplegic. J Cell Biochem 1997; 67(1):1-12.
103. Xia Y, Karin M. The control of cell motility and epithelial morphogenesis by Jun kinases. Trends Cell Biol 2004; 14(2):94-101.
104. Huang C, Rajfur Z, Borchers C et al. JNK phosphorylates paxillin and regulates cell migration. Nature 2003; 424(6945):219-223.
105. Lee YH, Giraud J, Davis RJ et al. c-Jun N-terminal kinase (JNK) mediates feedback inhibition of the insulin signaling cascade. J Biol Chem 2003; 278(5):2896-2902.
106. Basu A, You SA, Haldar S. Regulation of Bcl2 phosphorylation by stress response kinase pathway. Int J Oncol 2000; 16(3):497-500.
107. Fan M, Goodwin M, Vu T et al. Vinblastine-induced phosphorylation of Bcl-2 and Bcl-XL is mediated by JNK and occurs in parallel with inactivation of the Raf-1/MEK/ERK cascade. J Biol Chem 2000; 275(39):29980-29985.
108. Lei K, Davis RJ. JNK phosphorylation of Bim-related members of the Bcl2 family induces Bax-dependent apoptosis. Proc Natl Acad Sci USA 2003; 100(5):2432-2437.
109. She QB, Ma WY, Zhong S et al. Activation of JNK1, RSK2, and MSK1 is involved in serine 112 phosphorylation of Bad by ultraviolet B radiation. J Biol Chem 2002; 277(27):24039-24048.
110. Zhang Y, Zhou L, Miller CA. A splicing variant of a death domain protein that is regulated by a mitogen-activated kinase is a substrate for c-Jun N-terminal kinase in the human central nervous system. Proc Natl Acad Sci USA 1998; 95(5):2586-2591.
111. Chow CW, Dong C, Flavell RA et al. c-Jun NH(2)-terminal kinase inhibits targeting of the protein phosphatase calcineurin to NFATc1. Mol Cell Biol 2000; 20(14):5227-5234.
112. Gupta S, Campbell D, Derijard B et al. Transcription factor ATF2 regulation by the JNK signal transduction pathway. Science 1995; 267(5196):389-393.
113. Lim CP, Cao X. Serine phosphorylation and negative regulation of Stat3 by JNK. J Biol Chem 1999; 274(43):31055-31061.

114. Adler V, Pincus MR, Minamoto T et al. Conformation-dependent phosphorylation of p53. Proc Natl Acad Sci USA 1997; 94(5):1686-1691.
115. Whitmarsh AJ, Yang SH, Su MS et al. Role of p38 and JNK mitogen-activated protein kinases in the activation of ternary complex factors. Mol Cell Biol 1997; 17(5):2360-2371.
116. Ducret C, Maira SM, Lutz Y et al. The ternary complex factor Net contains two distinct elements that mediate different responses to MAP kinase signalling cascades. Oncogene 2000; 19(44):5063-5072.
117. Zhang Y, Zhong S, Dong Z et al. UVA induces Ser381 phosphorylation of p90RSK/MAPKAP-K1 via ERK and JNK pathways. J Biol Chem 2001; 276(18):14572-14580.
118. Le S, Connors TJ, Maroney AC. c-Jun N-terminal kinase specifically phosphorylates p66ShcA at serine 36 in response to ultraviolet irradiation. J Biol Chem 2001; 276(51):48332-48336.
119. Matsuda S, Miwa Y, Hirata Y et al. Negative feedback loop in T-cell activation through MAPK-catalyzed threonine phosphorylation of LAT. Embo J 2004; 23(13):2577-2585.
120. Wiltshire C, Matsushita M, Tsukada S et al. A new c-Jun N-terminal kinase (JNK)-interacting protein, Sab (SH3BP5), associates with mitochondria. Biochem J 2002; 367(Pt 3):577-585.
121. Goedert M, Hasegawa M, Jakes R et al. Phosphorylation of microtubule-associated protein tau by stress-activated protein kinases. FEBS Lett 1997; 409(1):57-62.
122. Dai R, Frejtag W, He B et al. c-Jun NH2-terminal kinase targeting and phosphorylation of heat shock factor-1 suppress its transcriptional activity. J Biol Chem 2000; 275(24):18210-18218.

CHAPTER 2

Regulation of JNK by Small GTPases

Maria Julia Marinissen* and J. Silvio Gutkind

The Ras Superfamily of Small GTPases

The *ras* oncogene was first isolated as the transforming sequence encoded by the genome of highly tumorigenic retroviruses.[1] Subsequently, activating mutations in the three human *ras* genes, H-*ras*, K-*ras* and N-*ras* were detected in many frequent human cancers, including those of the lung, colon, urinary bladder, gall bladder, and pancreas.[2] The cellular *ras* genes encode proteins of 21 kilodaltons that bind guanine nucleotides,[3,4] and are inactive when bound to GDP, but become activated upon exchange of this nucleotide for GTP when stimulated by guanine nucleotide exchange factors (GEFs). In turn, the GTP-bound forms of Ras proteins control the activity of their specific downstream effector molecules, until Ras signaling is terminated by the hydrolysis of GTP to GDP, a process that is accelerated by the action of GTPase activating proteins (GAPs) that stimulate the intrinsic GTPase activity of Ras.[5] Thus, Ras proteins function as regulated binary molecular switches, depending whether they are bound to GDP (off state) or GTP (on state), and their activity is tightly controlled by the coordinated regulation of GEFs and GAPs.[5] Additionally, for some members of the Ras superfamily their GDP-bound state can be stabilized by GDP-dissociation inhibitors (GDIs), which thus act as negative regulators (Fig. 1). Mutations in some key residues, such as single amino acid substitutions in codons 12, 13 or 61, can render Ras proteins resistant to the effects of GAPs, thus remaining persistently activated and unmasking their transforming potential.[2] K-Ras, H-Ras, and N-Ras are the prototypes of what is now known to be a large superfamily of small GTPases that includes over 100 members, which can be divided in eight distinct branches according to their sequence identity and protein function: Ras, Rab, Rho, Ran, Gem/Rad, Gie1/2, Sar, and Arf/Arl.[6,7] These small GTPases have been linked to a wide variety of cellular responses, ranging from vesicular protein trafficking by the Rab and Arf families to signal transduction by the Ras and Rho families.[6] In particular, the Ras family transduces proliferative signals from growth factors and their receptors to the regulation of genes involved in cell proliferation and differentiation in organisms as diverse as yeast, fly, and mammals. One of the most studied molecular events initiated by Ras is the activation of the p42 and p44 mitogen activated protein kinases (MAPKs), that are also known as extracellular regulated kinases 2 and 1 (ERK2 and ERK1), respectively.[8,9] In this pathway, Ras is the first molecule activated by a number of cell surface receptors, including tyrosine-kinase polypeptide growth factor receptors and G protein-coupled receptors (GPCRs), through Ras specific GEFs, such as Sos and Ras-GRF.[5,10,11] In the GTP bound form, Ras initiates the activity of a cascade of cytoplasmic kinases by binding and recruiting to the membrane the serine-threonine protein kinase c-Raf[12,13] or its related isoforms, A-Raf and B-Raf, which then phosphorylate the dual specificity MAPKKs, MEK1 and MEK2, which subsequently activate the MAPKs ERK1 and

*Corresponding Author: Maria Julia Marinissen—Instituto de Investigaciones Biomedicas A. Sols UAM-CSIC, Departamento de Bioquímica, Facultad de Medicina, Universidad Autonoma de Madrid, Arzobispo Morcillo 4, Madrid 28029, Spain. Email: mjmarinissen@iib.uam.es

The JNK Signaling Pathway, edited by Anning Lin. ©2006 Eurekah.com.

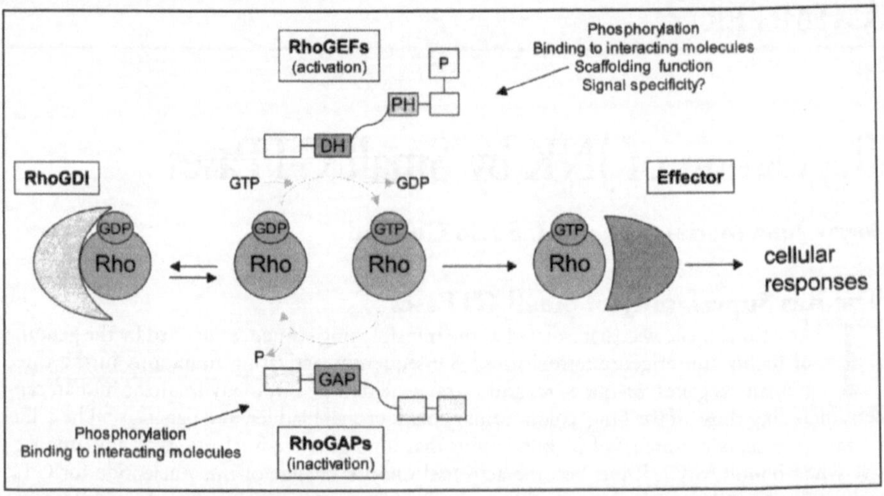

Figure 1. The Rho GTPases on/off switches. Transition between GDP-bound inactive and GTP-bound active states of Rho GTPases is regulated by the action of guanine nucleotide exchange factors (GEFs), GTPase activating proteins (GAPs) and GDP-dissociation inhibitors (GDIs). GEFs and GAPs can be regulated by phosphorylation or binding to scaffolding and regulatory molecules upon stimulation.

ERK2. These MAPKs translocate to the nucleus and phosphorylate transcription factors involved the regulation of growth promoting genes (Fig. 2) (for review see ref. 8).

The elucidation of the sequential assembling of the molecular components of the Ras-ERK cascade and the ultimate effects of ERK on nuclear events controlling gene expression represented one of the first examples of a molecularly defined pathway capable of transmitting mitogenic signals initiated at the level of the plasma membrane to the nucleus, thereby promoting normal and aberrant cell growth. Indeed, activated forms of Ras, Raf or even MEK were sufficient to induce a transformed phenotype in cultured cells, and oncogenic forms of Ras that can not bind Raf and consequently are unable to affect ERK1/2 activity, lacked transforming activity.[14] However, other effector domain mutants of Ras that activate the ERK1/2 pathway fully were defective in their transforming capacity, thus raising the possibility of that additional Ras effector molecules might be required for its proliferative response.[14]

In this regard, early in 1990, Kyriakis et al identified a protein kinase that was activated in liver in response to cyclohexamide exposure.[15] In the subsequent four years, independent efforts revealed that this kinase phosphorylates the transcription factor c-Jun,[16] that it belonged to the MAPK group of proline-targeted serine/threonine protein kinases[17] and that it can be activated by U.V. light and H-Ras.[18] Thus, this kinase, termed c-Jun N-terminal Kinase, JNK, or Stress Activated Protein Kinase, SAPK, represented a raising star in the ever evolving signal transduction galaxy. Given the fact that, like ERKs, JNK also translocates to the nucleus and phosphorylates transcription factors, it became almost a given to think that JNK might represent the "missing link" between Ras and nuclear responses, and that JNK was most likely activated by Ras through a kinase cascade similar to that leading to ERK activation. Soon after, however, the emerging experimental evidence started to suggest that this was not strictly the case.

Rac and Cdc42 Take the Center Stage

Unexpectedly, new studies showed that H-Ras did not activate JNK fully.[19] Moreover, several agonists know to act on the Ras-ERK pathway in certain cell types were ineffective to activate JNK.[17,20] Conversely, TNFα and IL-1 behaved as potent activators of JNK but had

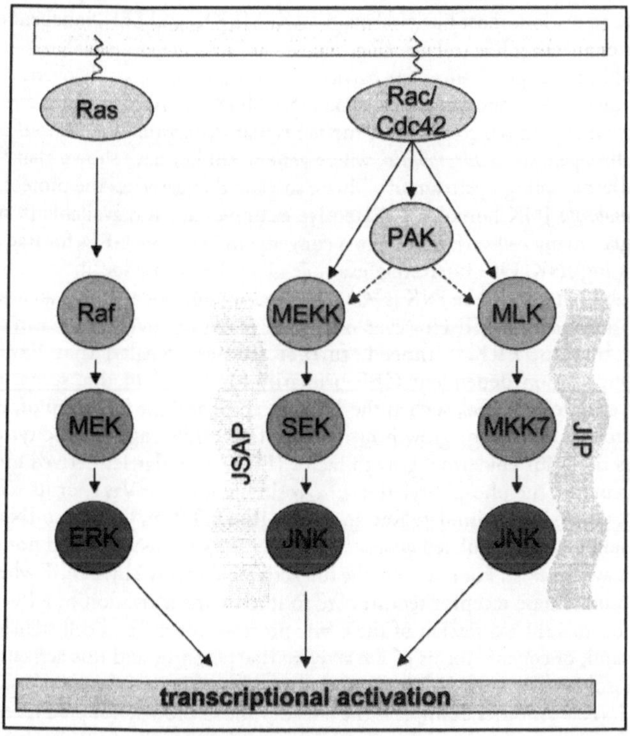

Figure 2. The Ras-ERK and the Rac/Cdc42 signaling modules. Activated Ras, by means of a cascade that activates Raf and MEK, induces the activation and nuclear translocation of ERK. Activated Rac and Cdc42, utilize different effectors such as PAK, MEKK, and MLKs to induce JNKKs and JNK. The signal specificity is provided by the nature of particular scaffolding proteins. Upon nuclear translocation, both ERKs and JNKs are capable of phosphorylating and activating transcription factors. Dashed arrows indicate putative interactions or activations.

only a poor effect on ERK1/2.[17] In this exciting scenario, the observation that Raf did not activated JNK whereas another MAPKKK, MEKK1, acted as a positive modulator of the pathway,[19] suggested that additional, yet to be identified signaling routes regulating JNK might also exist.

At that time, PAK, a kinase similar to Ste20p, a member of a defined MAPK pathway in yeast, was found to bind Rac and Cdc42, two members of the Rho family of small GTPases.[21] This finding prompted our group to explore whether these Rho proteins, up to that time believed to be primarily involved in the reorganization of the actin cytoskeleton,[22] were able to stimulate biochemical routes leading to JNK activation. Indeed, by the transient transfection of tagged JNK1 and ERK2 with activated mutants of Rac and Cdc42 in Cos-7 cells, we were able to show that these two GTPases activate JNK potently, but not ERK.[20] Similarly, expression of Ost and Dbl, two oncogenic GEFs for Rho proteins, was sufficient to trigger the strong activation of JNK without stimulating ERK, whereas dominant negative forms of Rac and Cdc42 blocked JNK activation by EGF. These results were further supported by the fact that RhoGDI and p190RhoGap, two molecules that inhibit Rac and Cdc42 activities, were capable of blocking the activation of JNK by Ost and Dbl.[20] Similar results were obtained by others using HeLa cells, who also observed that Rac stimulated the activity of SEK, a JNKK known to directly phosphorylate and activate JNK.[23] The emerging picture from these studies was that in mammalian cells, Rac and Cdc42 could initiate the activity of a linear signaling pathway leading to JNK activation, most likely through PAK, MEKK, and SEK, independently of the signaling route connecting Ras to ERK (Fig. 2).

Although all initial experiments were carried out in transient or stable overexpression systems, the key role for Rac and Cdc42 in signaling to JNK in vivo has now been extensively documented. A clear example comes from the molecular dissection of the dorsal closure process during embryogenesis in *Drosophila*, where genetic studies have shown that this mechanism requires dRac1, acting upstream of a kinase cascade that activates the protein product of *basket*, the *Drosophila* JNK homolog.[24] Extensive examples are also available in mammalian cells. In particular, many cell surface receptors converge to stimulate GEFs for Rac and Cdc42 thereby stimulating JNK. One interesting example of a GEF whose specificity was discovered by virtue of its ability to stimulate JNK is Vav1. Vav was initially identified as an oncoprotein[25] and thought to act as a Ras-GEF. However, oncogenic forms of Vav1 were shown to stimulate JNK potently but not ERK.[26] Indeed, further studies revealed that Vav1 acts as a tyrosine-phosphorylation dependent GEF primarily for Rac.[27] In this scenario, complex multimeric cell surface receptors, such as the T-cell receptor and the IgE receptor, that activate nonreceptor tyrosine kinases, or growth factor receptors exhibiting intrinsic tyrosine kinase activity, such as those for epidermal growth factor (EGF) and platelet-derived growth factor (PDGF), can promote the phosphorylation in tyrosine residues of Vav1, or its related GEFs, Vav2 and Vav3, which then stimulate Rac and other Rho GTPases leading to JNK activation and the expression of JNK-regulated genes.[28] Interplay between receptor and nonreceptor tyrosine kinases may also exist. For example, the full activation of Vav2 by PDGF when acting on its cognate tyrosine-kinase receptor requires Src to initiate the activation of a Rac-dependent pathway that controls the expression of the c-*myc* proto-oncogene and cell transformation.[29] On the other hand, oncogenic forms of Src activate Rac potently, and this activation is mediated by the phosphorylation of Vav2 and another unrelated Rac-GEF, Tiam, resulting in the stimulation of JNK.[30] Another example is the effect of the G protein-coupled receptor agonist endothelin A, which activates Src and induces the phosphorylation of FRG, a Cdc42-specific GEF, and that leads to the stimulation of JNK via Cdc42.[31]

Ras-Rac/Cdc42 Hierarchical Relationships and Guanine Nucleotide Exchange Factors (GEFs)

Whether Rac and Cdc42 act downstream from Ras has been a matter of great interest. As such, Rac is required for Ras-mediated cell transformation and epistatic analysis placed Rac as a key downstream target in Ras proliferative signaling (for review see ref. 32). However, the biochemical mechanism regulating the cross-talk among these small GTP-binding proteins remains to be fully elucidated. Regarding the pathway by which Ras participates in JNK activation, recent studies have begun to shed new light into this issue. Indeed, although tyrosine kinase receptors can cooperate with Src and Src-like kinases to stimulate Vav and consequently activate Rac and JNK, in some cases the function of Ras or Ras GEFs are required for the full activation of Rac by growth factors. For example, the EGF receptor, EGFR, can phosphorylate Eps8, which binds an adaptor protein known as E3b1/Abi-1. Eps8 and E3b1 can then form a complex with Sos that exhibits Rac GEF activity in addition to its best known Ras-GEF activity, thereby coordinating the activation of Ras and Rac by EGFR.[33] Sos can also gain the ability to stimulate Rac through its DH domain as a consequence of the potent activation of PI3K by Ras.[34] Alternatively, Ras can bind directly to a Rac GEF, Tiam-1, through a Ras-binding domain, thereby enhancing the accumulation of Rac-GTP and promoting membrane ruffling and JNK activation.[35] The emerging picture from these studies is that growth factor receptors may utilize multiple mechanisms to stimulate Rac, some of which require the coordinated activation of Ras. In turn, Ras may stimulate Rac activation by complementary mechanisms that include the direct activation of Rac GEFs, such as Tiam-1, and the concomitant stimulation of the Rac GEF activity of Sos and other Rac GEFs through the enhanced accumulation of the lipid products of PI3K activity.

The Pathway Linking Rac and Cdc42 to JNK

PAKs

Several groups have observed that members of the PAK family of serine/threonine protein kinase can mediate activation of JNK by Rac and/or Cdc42.[36-38] The mammalian PAKs have been divided into two groups. Group I comprises PAK1, 2 and 3, and they are structurally similar but differ with respect to their tissue distribution: *pak*1 is expressed in the brain, muscle, and spleen; *pak2* is ubiquitously expressed; and *pak*3 is expressed primarily in the brain. The group I is activated by Rac and Cdc42 upon binding to their N-terminal Cdc42 and Rac binding (CRIB) domain, which relieves the autoinhibitory activity of the N-terminus. After unfolding, PAK autophosphorylates on threonine 423 (for PAK1) within the activation loop of the kinase, and this phosphorylation prevents the refolding and consequent auto-inhibition, even in the absence of bound GTPase (for review see ref. 39). In turn, PAK regulates actin organization by inducing the formation of polarized filopodia and membrane ruffles,[40] similar to that provoked by the expression of Cdc42 and Rac, suggesting that PAK acts as an effector molecule in the pathway linking these small G proteins to the regulation of the actin-based cytoskeleton.

However, the regulatory mechanisms controlling localized PAK activity are likely more complex, reflecting the dynamic nature of the cytoskeletal changes that are required for cell motility. For example, the *Drosophila* homolog of PAK1, dPAK was found to colocalize with dynamic actin structures in focal adhesion complexes.[41] In line with this observation, mammalian PAK can be targeted to focal adhesion complexes through its interactions with the Rac exchange factor PIX, which indirectly binds FAK (focal adhesion kinase). Through these interactions, cell surface receptors that utilize FAK to transduce signals, such as integrins, can control localized Rac and PAK activity, and the cytoskeleton dynamics.[42] In addition, a particular Rac exchange factor, Trio, has been shown to regulate PAK activity during photoreceptor axon pathfinding in *Drosophila*.[43] Thus, a large body of experimental information suggests that PAK represents a key mediator of the ability of Rac and Cdc42 to promote cytoskeleton changes.[44,45] Whether JNK acts downstream of PAK in the pathway by which Rac and Cdc42 regulate the reorganization of the cellular actin is still unclear. For example, a recent publication shows that although dPAK is required for integrity of the cytoskeleton at the leading edge during dorsal closure, it does so by a mechanism that does not require JNK.[46]

The Group II PAKs (PAK 4, 5 and 6) are highly related to each other, exhibiting 75% similarity in their kinase domains, but are quite different from the Group I PAKs.[39] They also have a Rac/Cdc42 binding domain that is distinct from the CRIB found in the group I, and they lack an autoinhibitory region. Contrary to Group I PAKs, binding to activated GTPases does not stimulate their kinase activity in vitro, which indicates that the Group II PAKs are differently regulated. Although these kinases can mediate the effects of small GTPases on the cytoskeleton, more limited information is available on their activity towards the JNK pathway. For example, PAK4 binds Cdc42 preferentially and its overexpression induces actin polymerization and the formation of filopodia. However, although it activates JNK, this activation is somehow weaker than that provoked by Rac and Cdc42.[47] On the other hand, PAK5, a brain-specific PAK, promotes filopodia and neurite outgrowth and these biological effects appear not to be mediated by JNK.[48] Thus, the ability of small GTPases to promote cytoskeletal changes through Group I and Group II PAKs can be often dissociated from their stimulatory effect on JNK.

A more complex picture appears to emerge when analyzing the transforming activity of small GTPases. For example, Rac requires PAK to induce transformation in Rat-1 fibroblasts but not in NIH 3T3 cells[49] and a kinase-deficient mutant of PAK1 inhibits Ras transformation.[50] Moreover, Rac has been recently identified as one of the downstream effectors of the Kaposi's sarcoma-associated herpes virus GPCR during cellular transformation,[51] and PAK1 acts as a downstream target of Rac in the pathway by which this viral GPCR stimulates NFκB,[52] but not JNK (unpublished results). In this regard, although JNK activity has been linked to cell transformation (for review see ref. 53), its role as a target of Pak in this event is not clearly defined. For example, in Ras-induced Schwann cell transformation, dominant negative forms of PAK1 prevent this effect without affecting JNK activity.[54] Similarly, Rac3 can regulate the aberrant proliferation of breast cancer cells by a PAK-dependent pathway, but the expression of an interfering PAK1 fragment into a breast cancer cell line revealed that active Rac3 promotes PAK1 and JNK activities by two separate pathways. Only the Rac3-PAK pathway was critical for DNA synthesis and tumor growth independently of JNK.[55] Thus, studies on cell growth promotion by small GTPases suggest that PAK and JNK are not part of the same signaling route.

Aligned with this possibility, expression of individual GEFs in Cos-7 cells showed that their ability to activate PAK1 could be dissociated from their stimulatory activity on JNK, even though these GEFs act on Rac and/or Cdc42 to stimulate both pathways.[56] On the other hand, effector domain mutants of Rac and Cdc42 revealed that their ability to induce membrane ruffles and filipodia, respectively, involves residues that are required to interact with PAK, but Rac effector mutants that do not bind or activate PAK are nonetheless still able to stimulate JNK.[57] Together, these studies suggested that small GTPases can stimulate JNK by PAK-independent mechanisms.

MLKs

Consistent with the observation that PAK activates JNK only to a limited extent,[37,38] our group found that wild-type PAK does not cooperate with activated forms of Rac and Cdc42 to stimulate JNK.[58] In search for alternative candidates to mediate the activation of JNK by small GTPases, we used bioinformatic tools to identify additional CRIB-domain containing kinases, which led to the identification of mixed lineage kinase 3 (MLK3) as a putative candidate.[58] Further experiments demonstrated that MLK3 can act downstream of Rac and Cdc42 and upstream of SEK, also known as MKK4, thereby defining a linear kinase cascade by which these GTPases activate JNK.[58] Indeed, studies in fly have demonstrated that a similar cascade acts downstream from dRac in the dorsal closure process. This signaling route includes the MKKKK misshapen (Msn) and the *Drosophila* MLK homolog slipper (Slpr), which stimulate the dJNKK hemipterus (Hep) and subsequently dJNK, known as basket (Bsk) (for review see ref. 59). MLK3 belongs to a family of serine/threonine kinases that based on sequence similarities in their catalytic domain are clustered into three groups. The first group, the MLKs (MLK1-4), contains an amino-terminal src-homology 3 domain (SH3 domain), a kinase domain, a leucine zipper region, and a CRIB domain. The second group, the DLKs (dual-leucine-zippers-bearing kinases) that include LZK, exhibits a kinase domain followed by two leucine zipper. Finally, the ZAKs (zipper sterile-α-motif kinase) have both a leucine zipper and a sterile-α-motif (SAM) that is thought to mediate dimerization. All of them active JNK when overexpressed, and some of them can also act on p38 (for review see ref. 60). The CRIB domain present in the MLK group has been shown to bind Rac and Cdc42.[61] For MLK3, this interaction is thought to disrupt autoinhibitory interactions between MLK3 domains and to localize this kinase to specific subcellular compartments. Such is the case of MLK2 that colocalizes with JNK in punctuate structures along microtubules.[62]

The selectivity of MLKs for the JNK pathway may reside, in part, in the fact that MLK2, MLK3, DLK, and LZK have been all shown to associate with members of the JIP1 family of JNK binding proteins. Thus, JIP1-related proteins act as scaffolds organizing the assembly of specific kinase modules that act on each JNK isoform. Indeed, JIP1 binds MKK7 but not MKK4 (SEK), and JIP3 interacts with MLK3, MKK7, and JNK1/3, whereas its splice variant, JSAP, associates with MEKK, MKK4 and JNK1/2 (for review see ref. 53). Clearly, there appears to be a distinct specificity for each of these pathways, but how Rac and Cdc42 can choose to activate each of these effector modules upon stimulation is not known. In this regard, one likely mechanism determining signaling specificity may reside in the fact that some exchange factors may act themselves as scaffolds, thereby promoting the selective formation of particular signaling complexes. For example, Tiam1 interacts directly with JIP2, a scaffold for MLK3 and p38.[63] On the other hand, the ability to interact with these scaffolding molecules may also enable the activation of MLKs and JNKs without the requirement to activate Rho GTPases. As such, C3G, a GEF for the Ras-like GTPases Rap, binds JIP1 and activates JNK1 by a Ras and Rac-independent mechanism that involves MLKs.[64]

MEKKs

MEKK1 was one of the first kinases shown to activate the JNK pathway but not the ERK1/2 signaling route.[19] It was also observed later that MEKK1 can bind JNK[65] and that endogenous MEKK1 colocalizes with α-actinin in stress fibers and focal adhesions[66] in regions where also Rac and Cdc42 are found. All MEKKs, 1 to 4, regulate the JNK pathway but in contrast to MEKK2 and MEKK3, MEKK1 and MEKK4 exhibit structural features by which they can associate with Rac and Cdc42, as reflected by the observation that their kinase-inactive mutants can block the stimulation of JNK by Rac and Cdc42.[67] MEKKs activate JNK selectively through MKK4 (SEK) rather than MKK7 and this specificity appears to depend on the preferential binding to scaffold proteins, such as JSAP or filamin (for review see ref. 53). However, it is still not clear which MAPKKKs, MEKKs or MLKs, are relevant to each specific physiological stimulus. This uncertainty reflects the likely functional redundancy among these kinases, and the promiscuity of function often observed in overexpression and in vitro assays. In contrast, evidence for selectivity is now emerging from the systematic study of knockout animals and cells (for review see ref. 8). For example, MEKK1 -/- embryonic stem (ES) cells display normal activation of JNK in response to UV, heat shock, and anisomycin, but a dramatic decrease in JNK activity in response to nocodazole, serum, cold stress, and LPA.[68] Using a different targeting strategy for MEKK1, other groups found that this protein was necessary for the maximal JNK activation by growth factors, TNFα, IL-1, double-stranded RNA, and LPS.[69] Emerging technologies such as the use of RNA interference may soon help clarify the requirements for each particular MAPKKK in JNK activation by distinct stimuli, including small GTPases of the Rho family, in each cellular setting.

Other MAPKKKs

In addition to MLKs and MEKKs, several other MAPKKKs have been reported to activate the JNK signaling pathway. These include members of the ASK group (ASK1 and ASK2), TAK1, and TPL2. Interestingly, recent evidence suggests that ASK activates MKK4 by binding β-arrestin, and that this leads to the formation of a complex with JNK and its activation.[70] How Tpl2, also known as Cot, activates JNK is not known. Recent data indicates that Tpl2 forms a complex with hKSR-2 (kinase suppressor of Ras-2) and also coimmunoprecipitates with many signaling components of the Ras/MAPK pathway, including Ras, Raf, MEK-1, and ERK-1/2.[71] Whether a similar interaction with components of the JNK pathway exists is not yet known. Another MAPKKK, TAK1, has been characterized as a component of the TGF-β/bone morphogenetic protein signaling pathway.[72] TAK1 is also believed to be an important component of intracellular pathways involved in host responses to physiological and/or environmental stress signals during inflammation[73] through the NIK-IKK-NFκB cascade.[74] TAK1 participates in JNK activation as part of the ceramide-induced pro-apoptotic pathway[75] as well as in the pro-survival effect of IAP proteins.[76] Indeed, studies in flies suggest that dTAK can act as a JNKKK upstream of JNK in multiple contexts in the eye.[77,78] However, loss-of-function RNAi studies indicate that dTak is not strictly required for embryonic dorsal closure.[78] In a recent study, it was found that upon activation by IL-1, Ras becomes associated with IRAK, Traf-6, and TAK-1, leading to the assembly of a multiprotein signaling complex that is required for p38 MAPK activation, rather than JNK.[79] Thus, available information suggests that TAK may mediate the activation of p38, and in some cellular contexts that of JNK, but that this MAPKKK, as well as Ask1 and Tpl-2, are unlikely to participate in signaling from Rac and Cdc42 to JNK.

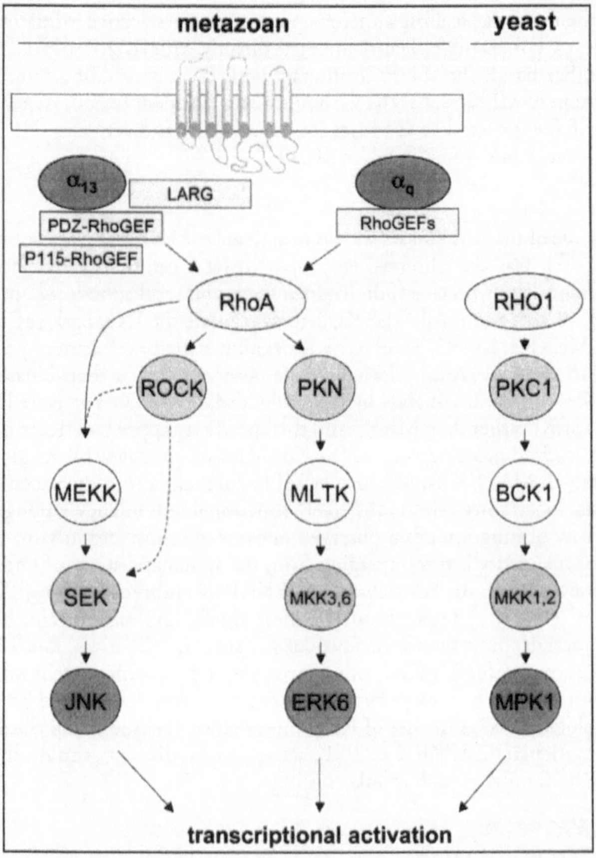

Figure 3. The Rho-JNK axis. G protein-coupled receptors can induce RhoA activity through RhoGEFs. RhoA induces JNK by a ROCK-dependent mechanism that involves SEK and can also bind MEKK directly. Whether ROCK activates SEK through MEKK is not known (dashed arrow). Through PKN and MKK3/6, RhoA activates the MAPK p38γ (ERK6) analogously to a yeast pathway whereby Rho1 acting on the PKN homolog PKC1, activates a p38 kinase cascade. The activation of these pathways results in the regulation of gene transcription and diverse biological responses.

A Role for Rho in JNK Activation

RhoA, RhoB and RhoC form a distinct cluster within the Rho family of GTPases, based on sequence similarity.[80] RhoA, the most extensively studied of them, is best known for its cytoskeletal effects, as expression of its activated mutants leads to the polymerization of actin into typical actin-based structures known as stress fibers (for review see ref. 32). In contrast to Rac and Cdc42, RhoA was found not to stimulate JNK or stimulated JNK poorly when coexpressed with epitope tagged forms of JNK in HeLa[23] or Cos-7 cells,[20] respectively, thus limiting the interest in exploring the possible contribution of RhoA to the activation of MAPKs in mammalian cells. However, in yeasts, Rho1p, a RhoA homolog, interacts directly with PKC1 to trigger a MAPK cascade involved in bud formation,[81] and in mammalian cells, RhoA was found to interact with PKN, a PKC-related protein (Fig. 3).[82] Based on these findings, we reevaluated whether RhoA could stimulate MAPKs. Indeed, we found that in other cell types, such as in HEK 293 cells, activated forms of RhoA, as well as RhoB and RhoC, could activated JNK, and that this activation was independent of PAK1.[58] In line with these findings, a

RhoA-specific exchange factor, mNET1, was shown to stimulate the JNK signaling pathways through RhoA.[83] As for the ability of Rac to stimulate JNK, emerging information suggests that the nature of the Rho GEF utilized to enhance Rho activity may have an impact on the signal output once Rho is stimulated. For example, JIP-1 was shown to bind directly the p190 RhoGEF in differentiated neurons,[84] thereby helping to assemble signaling complexes that likely include kinases acting on JNK in particular subcellular locations. On the other hand, the heterotrimeric G protein $G\alpha_{12}$ stimulates JNK potently,[85] and the pathway linking this G protein to JNK was found to involve the activation of Src tyrosine kinases and RhoA.[86]

In contrast, studies exploring the mechanism by which RhoA stimulate the expression from the c-Fos serum-response element (SRE) revealed that RhoA can enhance gene expression by activating the transcriptional activity of the Serum Response Factor (SRF) by a biochemical route that is independent from any known MAPK cascade.[87] Instead, RhoA activates the SRF by promoting actin polymerization through the activation of the serine/threonine kinase ROCK, which in turn phosporylates and activates a kinase known as LIMK that phosphorylates and inactivates the actin severing protein cofilin.[88] This favors the accumulation of polymerized actin, and the consequent decrease in the intracellular pool of monomeric actin, which appears to repress SRF function by binding a transcriptional coactivator known as MAL.[89] However, the possibility still exists that these two activities initiated by RhoA, actin polymerization and JNK activation, may be temporally and spatially coordinated.

In parallel studies addressing the mechanisms by which LPA controls the activity of transcription factors regulating the expression of c-Jun, we observed that LPA can stimulate RhoA to initiate the activity of a MAPK cascade resembling that previously shown in yeast. In this case, RhoA activates PKN, which indirectly stimulate MKK3/6 to activate the MAPK known as p38γ or ERK6 (Fig. 3). This MAPK phosphorylates the transcription factor MEF2 that binds a MEF2 response element within the *c-jun* promoter thereby promoting c-jun expression (Fig. 3).[90] Further work revealed the PKN itself forms a scaffold that binds and stimulates a kinase known as MLTK or MRK,[91] MKK3/6, and p38γ, thus providing the molecular basis for the specific activation of this particular p38 isoform.[92] In cardiac myocytes, RhoA can also stimulate p38α, albeit by a still unclear mechanism, and in turn p38α activates MEF2 leading to enhanced expression of atrial natriuretic factor (ANF) through GATA transcription factors.[93]

These findings and the likelihood that the expression of c-Fos and c-Jun could be coordinated to stimulate the transcriptional activity of AP-1 complexes, which are formed by Jun and Fos proteins, prompted us to examine whether ROCK, a kinase that acts as a downstream target for RhoA, also played a role in the regulation of c-Jun expression. To our surprise, we found that inhibition of ROCK diminished the expression of c-Jun in response to LPA or RhoA activation. Furthermore, we found that RhoA could activate JNK through ROCK in a variety of cell types, leading to the stimulation of the transcription factors ATF2 and c-Jun, which subsequently activate the *c-jun* promoter (Fig. 3).[94]

The finding that RhoA can regulate JNK may have broad implications in a number of physiological and pathological situations. For example, a recent study demonstrated that in renal cell carcinoma, hypoxic conditions promote the activation of the RhoA-JNK pathway, leading to the upregulation of the expression of the von Hippel-Lindau tumor-suppressor protein (pVHL).[95] A good example of cell type signal specificity and integration has been observed in keratinocytes. In these cells, matrix metalloproteinase-9 is induced in response to injury by a mechanism that involved the selective activation of p38 by Rac and Cdc42 and JNK by RhoA.[96] As for Rac, multiple kinases may link RhoA to JNK, depending on the cell type under investigation and the particular stimuli. As such, MEKK1 has been recently shown to bind RhoA and p115 RhoGEF, which results in enhanced MEKK1 activity.[97,98] Moreover, the regulation of JNK by p115RhoGEF and Rho appears to mediate the ability of $G\alpha_{13}$ to regulate primitive endoderm formation through a route that involved MEKK1 and MKK4.[99] Receptors coupled to $G\alpha_q$ and activated forms of this heterotrimeric G protein α subunit also activate RhoA[100] and JNK,[101] and a dominant negative form of Rho (RhoA N19) blocks JNK activation,[102] suggesting that $G\alpha_q$-coupled receptors may utilize RhoA to stimulate JNK. Although the specific Rho GEFs utilized by $G\alpha_q$ to activate Rho are still not fully defined, recent findings support a role for one particular Rho GEF, p63 RhoGEF, in this signaling route.[103]

The pathway from RhoA to JNK appears to be evolutionarily conserved. For example JNK has been shown to act downstream of RhoA in the pathway by which Frizzled regulates ommatidial polarity through Dishevelled during the morphogenesis of the eye in flies.[104] Similarly, in Xenopus laevis, paraxial protocadherin (XPAC) coordinates cell polarity during convergent extension initiated by the Wnt pathway through RhoA and JNK. Indeed, loss of XPAPC function blocks RhoA-mediated JNK activation and cell polarity during this highly coordinated developmental process.[105]

Others Small GTPases

Despite the central role for Rac, Cdc42, and Rho in the regulation of JNK, other related GTPases can also initiate the activity of signaling pathways leading to JNK activation. For example RhoG, which in spite of its name shares high sequence identity with Rac1 and Cdc42, interacts in a GTP-dependent manner with MLK3, thereby promoting JNK activation.[106] Interestingly, RhoG does not interact with other JNK activators such as PAKs.[106] However, RhoG can be directly activated by GEFs that also stimulate Rac and Cdc42, such as Vav2 and Dbl, arguing that RhoG function may be stimulated concomitantly with Rac and Cdc42 activation and that RhoG may utilize a subset of effectors in common with these GTPases.[106] TC10, which is also related to Rac and Cdc42, can stimulate the formation of long filopodia and JNK.[107] Another GTPase, Chp, which is highly related to Cdc42, was identified by a two-hybrid screen in search for PAK2-interacting molecules. Chp is a 36 kDa protein that exhibits 52% homology to Cdc42, and contains two additional sequences at its amino- and carboxyl-termini that are not found in other GTPases. The amino-terminus contains a poly-proline sequence, typically found in proteins binding Src homology 3 (SH3) domains, and its carboxyl-terminus appears to be important for PAK2 binding. By microinjecting Chp into cells, it has been observed that it is able to induce lamellipodia and to activate the JNK cascade.[108]

Wrch-1 (Wnt-1 responsive Cdc42 homolog), another GTPase highly related to Chp, whose mRNA level increases in response to Wnt-1 signaling in Wnt-1 transformed cells, shares 57% identity with Cdc42. As for Chp, also known as Wrch-2, Wrch-1 has a unique N-terminal domain that contains several putative PXXP SH3-binding motifs and can activate PAK-1 and JNK, induce filopodia formation and stress fiber dissolution, and stimulate quiescent cells to reenter the cell cycle phenocoping Wnt-1 in morphological transformation of mouse mammary epithelial cells.[109] Unlike Cdc42, Wrch-1 possesses an extremely rapid intrinsic guanine nucleotide exchange activity and N-terminal truncation of Wrch-1 enhances its ability to interact with and activate PAK and to cause growth transformation. Its N-terminus associates with the Grb2 SH3 domain-containing adaptor protein, and this association increases the levels of active Wrch-1 in cells, which suggests that Grb2 overcomes the negative regulation of the N-terminus of Wrch-1 thereby promoting its nucleotide exchange and ultimately JNK activation.[110] Thus, this atypical model of regulation of Wrch-1 broadens the repertoire of available mechanisms for JNK activation by small GTPases.

In addition to Rho GTPases, some Ras-related GTP-binding proteins can also activate JNK, albeit by a not fully elucidated mechanism. They include Ral, Rap2 and R-Ras. For Ral, expression of its activated mutant in *Drosophila* causes developmental defects that are genetically suppressed by loss of function mutations of hemipterous (dJNKK) and basket (dJNK).[111] Furthermore, activation of mammalian Ral by a Ral GEF, Rlf, can stimulate JNK potently, leading to the phsophorylation of the transactivation domain of c-Jun.[112] Of interest, Rlf is a downstream target of Ras[113] suggesting that Ral can participate in the activation of JNK by oncogenic forms of Ras. Furthermore, Ral appears to mediate the activation of JNK by H_2O_2, which results in the phosphorylation and nuclear translocation of the pro-apoptotic transcription factor FOXO4.[114] In the case of Rap2, in search for molecules mediating the activity of this Ras-related GTPase by a yeast two-hybrid screen, it was recently found that Rap2 can bind MAP4K4, a member of the STE20 group of protein kinases that are distantly related to PAK and MLK, which can stimulate the activity of JNK.[115] MAP4K4 was found to interact with Rap2 in a GTP-dependent fashion through its C-terminal citron homology domain but surprisingly, it does not interact with Rap1 or Ras.[115] Whether Rap2 mediates the activation of JNK by extracellular stimuli or cellular stress is at the present unknown. On the other hand, C3G, a GEF for Rap1 and R-Ras, has been shown to transduce the oncogenic signal form the v-Crk oncogene product to JNK.[116] Although initial studies focused on the possibility that Rap1 mediated JNK activation, recent work has revealed that R-Ras, but not Rap1, mediates JNK activation by v-Crk.[117] The nature of the molecules by which R-Ras activates JNK is still unknown. However, the R-Ras-dependent activation of JNK represents a critical component of the transforming pathway by which v-Crk acts.[117]

Conclusion

The realization that small GTP-binding proteins of the Rho family can stimulate JNK has provided a fascinating novel insight onto the biochemical routes by which numerous cell membrane receptors regulate nuclear events, thereby controlling normal and aberrant cell growth. Indeed, previously known primarily for their cytoskeletal effects, we now know that Rho proteins are an integral part of signaling pathways linking receptors and oncoproteins to the nucleus. As the complexity of the mechanisms by which these GTPases act began to unravel, we have also learned that the relative contribution of Rac, Cdc42 and RhoA and their downstream targets to the activation of JNK is likely to be cell type and stimulus-specific. In particular, the membrane receptors involved and the repertoire of available molecules in each cell type may determine the coupling specificity to each Rho GTPase and the selection of their effector molecules. In this regard, emerging evidence suggest that the nature of the GEF utilized can contribute to the specificity and duration of the activation of JNK and other MAPKs, as certain GEFs may favor the formation of specific molecular complexes based on their affinity and accessibility to scaffolding proteins, MAPKKKs, MAPKKs, and MAPKs. This recruitment and organization of distinct signaling complexes may ultimately dictate the choice of the intervening molecules and the nature and multiplicity of their possible interactions, thus providing the molecular basis for signal transduction specificity and the ability to integrate concomitant incoming signals, which may act in a combinatorial fashion. Although adding an unexpected layer of complexity to the study of signaling transduction through small GTPases, these mechanisms may ultimately ensure distinct biological outcomes in response to different extracellular cues. We can expect that ongoing high throughput efforts to systematically knock down proteins involved in signal transduction in model organisms and cells, combined with an expanding availability of genetically engineered animals lacking GEFs, GTPases, and their key downstream targets and regulatory molecules may soon help elucidate fully the complexity of the biologically relevant mechanism by which Rho GTPases stimulate nuclear events through the activation of JNKs and their related MAPKs.

References

1. Scolnick EM, Rands E, Williams D et al. Studies on the nucleic acid sequences of Kirsten sarcoma virus: A model for formation of a mammalian RNA-containing sarcoma virus. J Virol 1973; 12:458-63.
2. Bos JL. Ras oncogenes in human cancer: A review. Cancer Res 1989; 49:4682-9.
3. Barbacid M. Ras genes. Annu Rev Biochem 1987; 56:779-827.
4. Symons M, Takai Y. Ras GTPases: Singing in tune. Sci STKE 2001; (PE1 2001).
5. Bar-Sagi D, Hall A. Ras and Rho GTPases: A family reunion. Cell 2000; 103:227-38.
6. Etienne-Manneville S, Hall A. Rho GTPases in cell biology. Nature 2002; 420:629-35.
7. Okai T et al. Novel small GTPase subfamily capable of associating with tubulin is required for chromosome segregation. J Cell Sci 2004; 117:4705-15.
8. Pearson G et al. Mitogen-activated protein (MAP) kinase pathways: Regulation and physiological functions. Endocr Rev 2001; 22:153-83.

9. Whitmarsh AJ, Davis RJ. Signal transduction by MAP kinases: Regulation by phosphorylation-dependent switches. Sci STKE 1999, (PE1 1999).
10. Bonfini L, Karlovich CA, Dasgupta C et al. The Son of sevenless gene product: A putative activator of Ras. Science 1992; 255:603-6.
11. Wolfman A, Macara IG. A cytosolic protein catalyzes the release of GDP from p21ras. Science 1990; 248:67-9.
12. Vojtek AB, Hollenberg SM, Cooper JA. Mammalian Ras interacts directly with the serine/threonine kinase Raf. Cell 1993; 74:205-14.
13. Kerkhoff E, Rapp UR. The Ras-Raf relationship: An unfinished puzzle. Adv Enzyme Regul 2001; 41:261-7.
14. White MA et al. Multiple Ras functions can contribute to mammalian cell transformation. Cell 1995; 80:533-41.
15. Kyriakis JM, Avruch J. pp54 microtubule-associated protein 2 kinase. A novel serine/threonine protein kinase regulated by phosphorylation and stimulated by poly-L-lysine. J Biol Chem 1990; 265:17355-63.
16. Hibi M, Lin A, Smeal T et al. Identification of an oncoprotein- and UV-responsive protein kinase that binds and potentiates the c-Jun activation domain. Genes Dev 1993; 7:2135-48.
17. Kyriakis JM et al. The stress-activated protein kinase subfamily of c-Jun kinases. Nature 1994; 369:156-60.
18. Derijard B et al. JNK1: A protein kinase stimulated by UV light and Ha-Ras that binds and phosphorylates the c-Jun activation domain. Cell 1994; 76:1025-37.
19. Minden A et al. Differential activation of ERK and JNK mitogen-activated protein kinases by Raf-1 and MEKK. Science 1994; 266:1719-23.
20. Coso OA et al. The small GTP-binding proteins Rac1 and Cdc42 regulate the activity of the JNK/SAPK signaling pathway. Cell 1995; 81:1137-46.
21. Manser E, Leung T, Salihuddin H et al. A brain serine/threonine protein kinase activated by Cdc42 and Rac1. Nature 1994; 367:40-6.
22. Nobes C, Hall A. Regulation and function of the Rho subfamily of small GTPases. Curr Opin Genet Dev 1994; 4:77-81.
23. Minden A, Lin A, Claret FX et al. Selective activation of the JNK signaling cascade and c-Jun transcriptional activity by the small GTPases Rac and Cdc42Hs. Cell 1995; 81:1147-57.
24. Glise B, Noselli S. Coupling of Jun amino-terminal kinase and Decapentaplegic signaling pathways in Drosophila morphogenesis. Genes Dev 1997; 11:1738-47.
25. Katzav S, Martin-Zanca D, Barbacid M. vav, a novel human oncogene derived from a locus ubiquitously expressed in hematopoietic cells. Embo J 1989; 8:2283-90.
26. Crespo P et al. Rac-1 dependent stimulation of the JNK/SAPK signaling pathway by Vav. Oncogene 1996; 13:455-60.
27. Crespo P, Schuebel KE, Ostrom AA et al. Phosphotyrosine-dependent activation of Rac-1 GDP/GTP exchange by the vav proto-oncogene product. Nature 1997; 385:169-72.
28. Bustelo XR. Vav proteins, adaptors and cell signaling. Oncogene 2001; 20:6372-81.
29. Chiariello M, Marinissen MJ, Gutkind JS. Regulation of c-myc expression by PDGF through Rho GTPases. Nat Cell Biol 2001; 3:580-6.
30. Servitja JM, Marinissen MJ, Sodhi A et al. Rac1 function is required for Src-induced transformation. Evidence of a role for Tiam1 and Vav2 in Rac activation by Src. J Biol Chem 2003; 278:34339-46.
31. Miyamoto Y, Yamauchi J, Itoh H. Src kinase regulates the activation of a novel FGD-1-related Cdc42 guanine nucleotide exchange factor in the signaling pathway from the endothelin A receptor to JNK. J Biol Chem 2003; 278:29890-900.
32. Jaffe AB, Hall A. Rho GTPases in transformation and metastasis. Adv Cancer Res 2002; 84:57-80.
33. Scita G et al. EPS8 and E3B1 transduce signals from Ras to Rac. Nature 1999; 401:290-3.
34. Nimnual AS, Yatsula BA, Bar-Sagi D. Coupling of Ras and Rac guanosine triphosphatases through the Ras exchanger Sos. Science 1998; 279:560-3.
35. Lambert JM. et al. Tiam1 mediates Ras activation of Rac by a PI(3)K-independent mechanism. Nat Cell Biol 2002; 4:621-5.
36. Zhang S et al. Rho family GTPases regulate p38 mitogen-activated protein kinase through the downstream mediator Pak1. J Biol Chem 1995; 270:23934-6.
37. Bagrodia S, Derijard B, Davis RJ et al. Cdc42 and PAK-mediated signaling leads to Jun kinase and p38 mitogen-activated protein kinase activation. J Biol Chem 1995; 270:27995-8.
38. Frost JA, Xu S, Hutchison MR et al. Actions of Rho family small G proteins and p21-activated protein kinases on mitogen-activated protein kinase family members. Mol Cell Biol 1996; 16:3707-13.

39. Jaffer ZM, Chernoff J. p21-activated kinases: Three more join the Pak. Int J Biochem Cell Biol 2002; 34:713-7.
40. Sells MA et al. Human p21-activated kinase (Pak1) regulates actin organization in mammalian cells. Curr Biol 1997; 7:202-10.
41. Harden N et al. A Drosophila homolog of the Rac- and Cdc42-activated serine/threonine kinase PAK is a potential focal adhesion and focal complex protein that colocalizes with dynamic actin structures. Mol Cell Biol 1996; 16:1896-908.
42. Zhao ZS, Manser E, Loo TH et al. Coupling of PAK-interacting exchange factor PIX to GIT1 promotes focal complex disassembly. Mol Cell Biol 2000; 20:6354-63.
43. Newsome TP et al. Trio combines with dock to regulate Pak activity during photoreceptor axon pathfinding in Drosophila. Cell 2000; 101:283-94.
44. Bokoch GM. Regulation of cell function by Rho family GTPases. Immunol Res 2000; 21:139-48.
45. Bagrodia S, Cerione RA. Pak to the future. Trends Cell Biol 1999; 9:350-5.
46. Conder R et al. dPak is required for integrity of the leading edge cytoskeleton during Drosophila dorsal closure but does not signal through the JNK cascade. Dev Biol 2004; 276:378-90.
47. Abo A et al. PAK4, a novel effector for Cdc42Hs, is implicated in the reorganization of the actin cytoskeleton and in the formation of filopodia. Embo J 1998; 17:6527-40.
48. Dan C, Nath N, Liberto M et al. PAK5, a new brain-specific kinase, promotes neurite outgrowth in N1E-115 cells. Mol Cell Biol 2002; 22:567-77.
49. Tang Y, Yu J, Field J. Signals from the Ras, Rac, and Rho GTPases converge on the Pak protein kinase in Rat-1 fibroblasts. Mol Cell Biol 1999; 19:1881-91.
50. Tang Y et al. Kinase-deficient Pak1 mutants inhibit Ras transformation of Rat-1 fibroblasts. Mol Cell Biol 1997; 17:4454-64.
51. Montaner S et al. The small GTPase Rac1 links the Kaposi sarcoma-associated herpesvirus vGPCR to cytokine secretion and paracrine neoplasia. Blood 2004; 104:2903-11.
52. Dadke D, Fryer BH, Golemis EA et al. Activation of p21-activated kinase 1-nuclear factor kappaB signaling by Kaposi's sarcoma-associated herpes virus G protein-coupled receptor during cellular transformation. Cancer Res 2003; 63:8837-47.
53. Manning AM, Davis RJ. Targeting JNK for therapeutic benefit: From junk to gold? Nat Rev Drug Discov 2003; 2:554-65.
54. Tang Y et al. A role for Pak protein kinases in Schwann cell transformation. Proc Natl Acad Sci USA 1998; 95:5139-44.
55. Mira JP, Benard V, Groffen J et al. Endogenous, hyperactive Rac3 controls proliferation of breast cancer cells by a p21-activated kinase-dependent pathway. Proc Natl Acad Sci USA 2000; 97:185-9.
56. Zhou K. Guanine nucleotide exchange factors regulate specificity of downstream signaling from Rac and Cdc42. J Biol Chem 1998; 273:16782-6.
57. Westwick JK et al. Rac regulation of transformation, gene expression, and actin organization by multiple, PAK-independent pathways. Mol Cell Biol 1997; 17:1324-35.
58. Teramoto H et al. The small GTP-binding protein rho activates c-Jun N-terminal kinases/stress-activated protein kinases in human kidney 293T cells. Evidence for a Pak-independent signaling pathway. J Biol Chem 1996; 271:25731-4.
59. Xia Y, Karin M. The control of cell motility and epithelial morphogenesis by Jun kinases. Trends Cell Biol 2004; 14:94-101.
60. Gallo KA, Johnson GL. Mixed-lineage kinase control of JNK and p38 MAPK pathways. Nat Rev Mol Cell Biol 2002; 3:663-72.
61. Burbelo PD, Drechsel D, Hall A. A conserved binding motif defines numerous candidate target proteins for both Cdc42 and Rac GTPases. J Biol Chem 1995; 270:29071-4.
62. Nagata K et al. The MAP kinase kinase kinase MLK2 colocalizes with activated JNK along microtubules and associates with kinesin superfamily motor KIF3. Embo J 1998; 17:149-58.
63. Buchsbaum RJ, Connolly BA, Feig LA. Interaction of Rac exchange factors Tiam1 and Ras-GRF1 with a scaffold for the p38 mitogen-activated protein kinase cascade. Mol Cell Biol 2002; 22:4073-85.
64. Tanaka S, Hanafusa H. Guanine-nucleotide exchange protein C3G activates JNK1 by a ras-independent mechanism. JNK1 activation inhibited by kinase negative forms of MLK3 and DLK mixed lineage kinases. J Biol Chem 1998; 273:1281-4.
65. Xu S, Cobb MH. MEKK1 binds directly to the c-Jun N-terminal kinases/stress-activated protein kinases. J Biol Chem 1997; 272:32056-60.
66. Christerson LB, Vanderbilt CA, Cobb MH. MEKK1 interacts with alpha-actinin and localizes to stress fibers and focal adhesions. Cell Motil Cytoskeleton 19991; 43:186-98.
67. Fanger GR, Johnson NL, Johnson GL. MEK kinases are regulated by EGF and selectively interact with Rac/Cdc42. Embo J 1997; 16:4961-72.

68. Yujiri T, Sather S, Fanger GR et al. Role of MEKK1 in cell survival and activation of JNK and ERK pathways defined by targeted gene disruption. Science 1998; 282:1911-4.
69. Xia Y et al. MEK kinase 1 is critically required for c-Jun N-terminal kinase activation by proinflammatory stimuli and growth factor-induced cell migration. Proc Natl Acad Sci USA 2000; 97:5243-8.
70. Miller WE et al. Identification of a motif in the carboxyl terminus of beta -arrestin2 responsible for activation of JNK3. J Biol Chem 2001; 276:27770-7.
71. Channavajhala PL et al. Identification of a novel human kinase supporter of Ras (hKSR-2) that functions as a negative regulator of Cot (Tpl2) signaling. J Biol Chem 2003; 278:47089-97.
72. Yamaguchi K et al. Identification of a member of the MAPKKK family as a potential mediator of TGF-beta signal transduction. Science 1995; 270:2008-11.
73. Lee J, Mira-Arbibe L, Ulevitch RJ. TAK1 regulates multiple protein kinase cascades activated by bacterial lipopolysaccharide. J Leukoc Biol 2000; 68:909-15.
74. Ninomiya-Tsuji J et al. The kinase TAK1 can activate the NIK-I kappaB as well as the MAP kinase cascade in the IL-1 signalling pathway. Nature 1999; 398252-6.
75. Shirakabe K et al. TAK1 mediates the ceramide signaling to stress-activated protein kinase/c-Jun N-terminal kinase. J Biol Chem 1997; 272:8141-4.
76. Sanna MG et al. IAP suppression of apoptosis involves distinct mechanisms: The TAK1/JNK1 signaling cascade and caspase inhibition. Mol Cell Biol 2002; 22:1754-66.
77. Takatsu Y et al. TAK1 participates in c-Jun N-terminal kinase signaling during Drosophila development. Mol Cell Biol 2000; 20:3015-26.
78. Mihaly J et al. The role of the Drosophila TAK homologue dTAK during development. Mech Dev 2001; 102:67-79.
79. McDermott EP, O'Neill LA. Ras participates in the activation of p38 MAPK by interleukin-1 by associating with IRAK, IRAK2, TRAF6, and TAK-1. J Biol Chem 2002; 277:7808-15.
80. Coleman ML, Marshall CJ, Olson MF. RAS and RHO GTPases in G1-phase cell-cycle regulation. Nat Rev Mol Cell Biol 2004; 5:355-66.
81. Nonaka H et al. A downstream target of RHO1 small GTP-binding protein is PKC1, a homolog of protein kinase C, which leads to activation of the MAP kinase cascade in Saccharomyces cerevisiae. Embo J 1995; 14:5931-8.
82. Watanabe G et al. Protein kinase N (PKN) and PKN-related protein rhophilin as targets of small GTPase Rho. Science 1996; 271;645-8.
83. Alberts AS, Treisman R. Activation of RhoA and SAPK/JNK signalling pathways by the RhoA-specific exchange factor mNET1. Embo J 1998; 17:4075-85.
84. Meyer D, Liu A, Margolis B. Interaction of c-Jun amino-terminal kinase interacting protein-1 with p190 rhoGEF and its localization in differentiated neurons. J Biol Chem 1999; 274:35113-8.
85. Prasad MV, Dermott JM, Heasley LE et al. Activation of Jun kinase/stress-activated protein kinase by GTPase-deficient mutants of G alpha 12 and G alpha 13. J Biol Chem 1995; 270:18655-9.
86. Nagao M, Kaziro Y, Itoh H. The Src family tyrosine kinase is involved in Rho-dependent activation of c-Jun N-terminal kinase by Galpha12. Oncogene 1999; 18:4425-34.
87. Hill CS, Wynne J, Treisman R. The Rho family GTPases RhoA, Rac1, and CDC42Hs regulate transcriptional activation by SRF. Cell 1995; 81:1159-70.
88. Sotiropoulos A, Gineitis D, Copeland J et al. Signal-regulated activation of serum response factor is mediated by changes in actin dynamics. Cell 1999; 98:159-69.
89. Miralles F, Posern G, Zaromytidou AI et al. Actin dynamics control SRF activity by regulation of its coactivator MAL. Cell 2003; 113:329-42.
90. Marinissen MJ, Chiariello M Gutkind JS. Regulation of gene expression by the small GTPase Rho through the ERK6 (p38 gamma) MAP kinase pathway. Genes Dev 2001; 15:535-53.
91. Gotoh I, Adachi M, Nishida E. Identification and characterization of a novel MAP kinase kinase kinase, MLTK. J Biol Chem 2001; 276:4276-86.
92. Takahashi M et al. Regulation of a mitogen-activated protein kinase kinase kinase, MLTK by PKN. J Biochem (Tokyo) 2003; 133:181-7.
93. Charron F et al. Tissue-specific GATA factors are transcriptional effectors of the small GTPase RhoA. Genes Dev 2001; 15:2702-19.
94. Marinissen MJ et al. The small GTP-binding protein RhoA regulates c-jun by a ROCK-JNK signaling axis. Mol Cell 2004; 14:29-41.
95. Turcotte S, Desrosiers RR, Beliveau R. Hypoxia upregulates von Hippel-Lindau tumor-suppressor protein through RhoA-dependent activity in renal cell carcinoma. Am J Physiol Renal Physiol 2004; 286:F338-48.

96. Turchi L et al. Cross-talk between RhoGTPases and stress activated kinases for matrix metalloproteinase-9 induction in response to keratinocytes injury. J Invest Dermatol 2003; 121:1291-300.
97. Gallagher ED, Gutowski S, Sternweis PC et al. RhoA binds to the amino terminus of MEKK1 and regulates its kinase activity. J Biol Chem 2004; 279:1872-7.
98. Christerson LB et al. p115 Rho GTPase activating protein interacts with MEKK1. J Cell Physiol 2002; 192:200-8.
99. Lee YN, Malbon CC. Wang HY. G alpha 13 signals via p115RhoGEF cascades regulating JNK1 and primitive endoderm formation. J Biol Chem 2004; 279:54896-904.
100. Chikumi H, Vazquez-Prado J, Servitja JM et al. Potent activation of RhoA by Galpha q and Gq-coupled receptors. J Biol Chem 2002; 277:27130-4.
101. Marinissen MJ, Servitja JM, Offermanns S et al. Thrombin protease-activated receptor-1 signals through Gq- and G13-initiated MAPK cascades regulating c-Jun expression to induce cell transformation. J Biol Chem 2003; 278:46814-25.
102. Yamauchi J et al. Galphaq-dependent activation of mitogen-activated protein kinase kinase 4/c-Jun N-terminal kinase cascade. Biochem Biophys Res Commun 2001; 288:1087-94.
103. Lutz S et al. The guanine nucleotide exchange factor p63RhoGEF - a specific link between Gq/11-coupled receptor signaling and RhoA. J Biol Chem 2005.
104. Strutt DI, Weber U, Mlodzik M. The role of RhoA in tissue polarity and Frizzled signalling. Nature 1997; 387:292-5.
105. Unterseher F et al. Paraxial protocadherin coordinates cell polarity during convergent extension via Rho A and JNK. Embo J 2004; 23:3259-69.
106. Wennerberg K et al. RhoG signals in parallel with Rac1 and Cdc42. J Biol Chem 2002; 277:47810-7.
107. Murphy GA et al. Cellular functions of TC10, a Rho family GTPase: Regulation of morphology, signal transduction and cell growth. Oncogene 1999; 18:3831-45
108. Aronheim A et al. Chp, a homologue of the GTPase Cdc42Hs, activates the JNK pathway and is implicated in reorganizing the actin cytoskeleton. Curr Biol 1998; 8:1125-8.
109. Tao W, Pennica D, Xu L et al. Wrch-1, a novel member of the Rho gene family that is regulated by Wnt-1. Genes Dev 2001; 15:1796-807.
110. Shutes A, Berzat AC, Cox AD et al. Atypical mechanism of regulation of the Wrch-1 Rho family small GTPase. Curr Biol 2004; 14:2052-6.
111. Sawamoto K et al. The Drosophila Ral GTPase regulates developmental cell shape changes through the Jun NH(2)-terminal kinase pathway. J Cell Biol 1999; 146:361-72.
112. de Ruiter ND, Wolthuis RM, van Dam H et al. Ras-dependent regulation of c-Jun phosphorylation is mediated by the Ral guanine nucleotide exchange factor-Ral pathway. Mol Cell Biol 2000; 20:8480-8.
113. O'Gara MJ, Zhang X, Baker L et al. Characterization of the Ras binding domain of the RalGDS-related protein, RLF. Biochem Biophys Res Commun 1997; 238:425-9.
114. Essers MA et al. FOXO transcription factor activation by oxidative stress mediated by the small GTPase Ral and JNK. Embo J 2004; 23:4802-12.
115. Machida N et al. Mitogen-activated protein kinase kinase kinase kinase 4 as a putative effector of Rap2 to activate the c-Jun N-terminal kinase. J Biol Chem 2004; 279:15711-4.
116. Tanaka S, Ouchi T, Hanafusa H. Downstream of Crk adaptor signaling pathway: Activation of Jun kinase by v-Crk through the guanine nucleotide exchange protein C3G. Proc Natl Acad Sci USA 1997; 94:2356-61.
117. Mochizuki N et al. Crk activation of JNK via C3G and R-Ras. J Biol Chem 2000; 275:12667-71.

Chapter 3

Physiological Roles of the Different JNKs:

Lessons from the JNK Knockouts

Kanaga Sabapathy*

Abstract

Structural and functional similarities are frequently found among many proteins, and such resemblance often suggests common modes of action or regulation. However, the question that often remains unresolved is if such similar molecules are evolutionarily conserved to perform exactly similar and redundant functions, or if they have specific and different biological roles that have yet to be uncovered. One such family of proteins is the c-Jun-N-terminal (JNK) kinases that are present ubiquitously in the organism. The JNK subgroup consists of three members—JNK1, JNK2 and JNK3—which are highly homologous and are present as multiple isoforms. Although combinatorial use of the various JNKs and their upstream kinases are thought to lead to differential regulation of various cellular processes, the specific roles of the JNK proteins were not clarified till the generation of the JNK mutant mice. This review will thus attempt to summarize the specific and redundant roles of JNK proteins, by analyzing the data obtained using the various JNK knockout mice.

Introduction

The c-Jun-amino-terminal kinase (JNK) was identified and purified in the early 1990s as the protein kinase activity that was induced by cyclohexamide, and as the one that induced c-Jun phosphorylation upon UV irradiation.[1,2] Subsequent cloning efforts revealed the presence of three distinct JNK genes encoding at least ten isoforms.[3-6] The three *Jnk* genes share a very high degree of homology and are able to phosphorylate substrates such as c-Jun and ATF-2 to varying extents in vitro kinase assays.[6] Hitherto, most in vitro studies have used c-Jun as the model substrate to investigate the nature of JNK-substrate binding activity and specificity. All three JNKs have been shown to bind to the delta domain of c-Jun and phosphorylate it on serines 63 and 73, leading to its activation.[4] Of the two, JNK2 has been shown to have a twenty-five fold higher binding affinity for c-Jun than JNK1.[4] However, many studies employing transient transfection experiments indicated that expression of any one of the *Jnk* genes resulted in the phosphorylation of substrates, leading to the speculation that the *Jnk* genes are redundant and play overlapping and probably similar roles. Meanwhile, it was established that JNK activity was activated in various cellular processes and was found to be important in transducing stress signals, such as those leading to apoptosis induction.[7] However, the second wave of studies trying to ascertain the role of the *Jnk* genes in apoptosis led to controversial results. Overexpression of the JNK genes led to pro-apoptotic effects in some situations and anti-apoptotic effects in others.[8-10] This led to the speculation that the JNKs may have different

*Kanaga Sabapathy—Laboratory of Molecular Carcinogenesis, Division of Cellular and Molecular Research, National Cancer Centre, 11 Hospital Drive, Singapore 169610 and the Department of Biochemistry, National University of Singapore, Singapore.
Email: cmrksb@nccs.com.sg

The JNK Signaling Pathway, edited by Anning Lin. ©2006 Eurekah.com.

properties in different cellular systems and perhaps depending also on the stimulating signal. Nonetheless, the physiological role and the reason for the evolutionary conservation of such highly similar genes were not addressed till the generation of the individual JNK knockout mice by gene targeting in the late 1990s. This review will attempt to summarize the genetic data obtained with the various JNK knockout mice, thereby highlighting the specific or combined roles of the various *Jnk* gene products in various physiological processes. Specificity of JNK substrates and regulation of JNK-mediated gene regulation are covered in the other reviews in this book.

General Overview of the JNK Knockouts

Mice lacking JNK1, JNK2 or JNK3 were generated by two separate groups of investigators.[11-15] Although expression analysis had indicated that JNK1 and JNK2 were expressed ubiquitously whereas JNK3 expression was restricted to the adult brain and heart regions,[5,16] it was evident that lack of any one *Jnk* gene did not compromise embryonic development, as the *Jnk1*$^{-/-}$, *Jnk2*$^{-/-}$ and *Jnk3*$^{-/-}$ mice were viable and fertile.[11-15] Overtly, none of them showed any major defects, except that the *Jnk1*$^{-/-}$ mutants were about 20-30% smaller in size at the early phases of their growth (Sabapathy K and Wagner EF, unpublished data). Nonetheless, apparent normality of life in the absence of the *Jnk* genes suggested that absence of one of the *Jnk* genes may be compensated for by the other *Jnk* genes, indicating functional redundancy among the *Jnk* genes during embryonic development.

Effects of JNK1 and JNK2 Deficiency in Mice

Though individual JNK knockouts appeared to develop normally, absence of both JNK1 and JNK2 resulted in embryonic lethality.[17,18] This lethal phenotype is specific to combined loss of JNK1 and JNK2 alone, as loss of JNK2 and JNK3 or JNK1 and JNK3 did not result in lethality,[18] indicating that JNK3 has no essential role during embryonic development. By contrast, JNK1 and JNK2 are essential components of the signaling cascade regulating development processes and appear to play overlapping and similar roles that cannot be replaced by JNK3 (Fig. 1). Hence, absence of JNK1 was replaceable by JNK2 and *vice versa*, but deletion of both was found to be crucial for development (Fig. 1).

Besides embryonic development, JNK1 and JNK2 were found to required for proper neural tube closure and brain development.[17,18] *Jnk1*$^{-/-}$*Jnk2*$^{-/-}$ embryos which die around embryonic day 11 (E11.0), exhibited an open neural tube (exencephaly) at the hindbrain level. Apoptosis was found to be reduced in the hindbrain neuroepithelium at E9.25 in these mutant embryos.[17] In contrast, a dramatic increase in cell death was observed one day later at E10.5 in both the hindbrain and forebrain regions, indicating that the JNKs play a pro-apoptotic role at earlier stages of neuronal development and subsequently an anti-apoptotic role at the later stages of brain development.[17] These studies for the first time demonstrated that the JNKs have both a pro- and anti-apoptotic roles in vivo, depending on the cellular context. However, the defects in brain development appear not be the cause of embryonic lethality as some *Jnk1*$^{-/-}$*Jnk2*$^{+/-}$ embryos, but not *Jnk1*$^{+/-}$*Jnk2*$^{-/-}$ embryos, displayed the exencephalic phenotype and were viable up to E16.0,[17] and suggested that the JNKs probably regulate other essential functions which have yet to be explored.

The phenotypic manifestation in *Jnk1*$^{-/-}$*Jnk2*$^{+/-}$ embryos argued that gene dosage might play a critical role in controlling JNK protein levels in the brain. Immunoblot analysis indicated that the total levels of JNK proteins were barely detectable in *Jnk1*$^{-/-}$*Jnk2*$^{+/-}$ brain extracts.[17] By contrast, *Jnk1*$^{+/-}$*Jnk2*$^{-/-}$ brain extracts contained considerable amounts of JNK proteins. Moreover, *Jnk1*$^{-/-}$*Jnk2*$^{+/-}$ spleens contained higher levels of JNK proteins than brain extracts of the same mice, suggesting that the reduction in JNK expression is specific to the brains of *Jnk1*$^{-/-}$*Jnk2*$^{+/-}$ mice.[17] This results also raised the possibility that although the essential functions of JNK1 and JNK2 may be similar and redundant, a critical threshold of JNK proteins are required for proper JNK signaling and hence, organismal development (Fig. 2). Consistent with this hypothesis, recent findings also indicated that *Jnk1*$^{-/-}$*Jnk2*$^{+/-}$ embryos exhibit defects

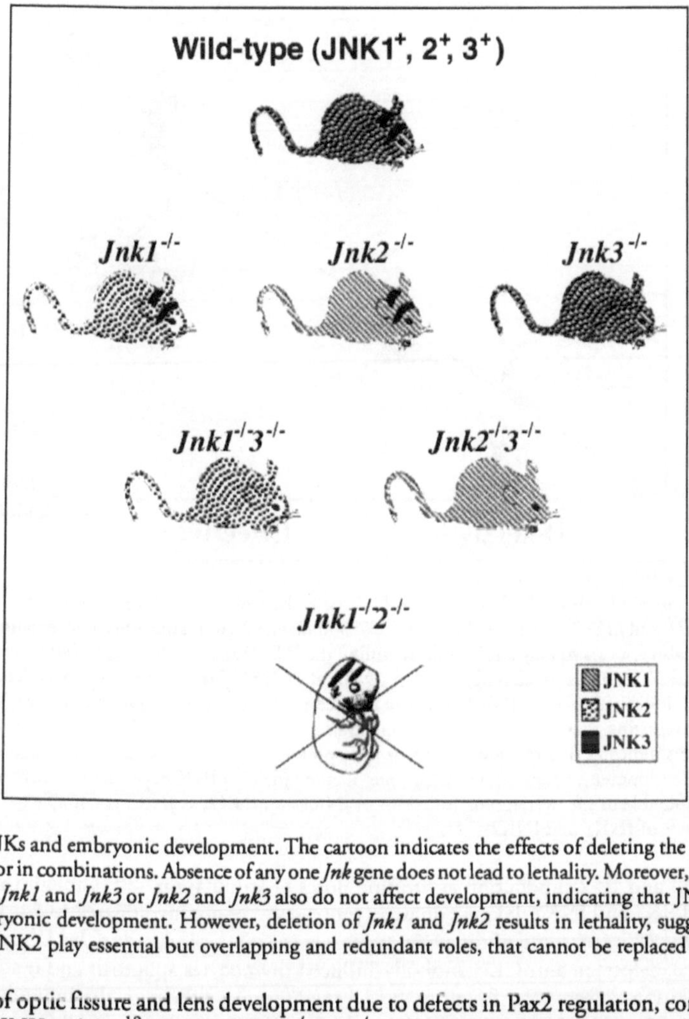

Figure 1. JNKs and embryonic development. The cartoon indicates the effects of deleting the various *Jnk* genes alone or in combinations. Absence of any one *Jnk* gene does not lead to lethality. Moreover, compound deletions of *Jnk1* and *Jnk3* or *Jnk2* and *Jnk3* also do not affect development, indicating that JNK3 has no role in embryonic development. However, deletion of *Jnk1* and *Jnk2* results in lethality, suggesting that JNK1 and JNK2 play essential but overlapping and redundant roles, that cannot be replaced by JNK3.

in closure of optic fissure and lens development due to defects in Pax2 regulation, concomitant to reduced JNK activity.[19] Moreover, $Jnk1^{-/-}Jnk2^{+/-}$ embryos were also shown to display delayed development in the epidermis, intestines and lungs due to decreased expression of epidermal growth factor.[20] These findings therefore suggest that critical levels of JNK proteins (either JNK1 or JNK2) are required for efficient JNK signaling during embryonic development, below which phenotypic defects are manifested (Fig. 2).

Similar but Nonredundant Roles of JNK1 and JNK2

Albeit JNK1 and JNK2 perform similar and redundant functions during embryonic development, they also perform similar but nonredundant functions in the adult mice. Essentially, similar phenotypes have been observed in several instances in both $Jnk1^{-/-}$ and $Jnk2^{-/-}$ mice. The lack of JNK1 or JNK2 resulted in resistance of thymocytes to anti-CD3 antibody-induce apoptosis in vitro and in vivo.[11,12] Moreover, concanavalin-A-induced hepatocyte apoptosis and consequent liver damage, which occurs through cell-bound TNFα, was considerably reduced in both $Jnk1^{-/-}$ and $Jnk2^{-/-}$ mice.[21] Furthermore, hyperoxia-induced lung injury and mortality was elevated similarly in $Jnk1^{-/-}$ and $Jnk2^{-/-}$ mice compared to wild-type controls.[22] Besides apoptosis, lack of either JNK1 or JNK2 also resulted in reduced mature T (total) cell

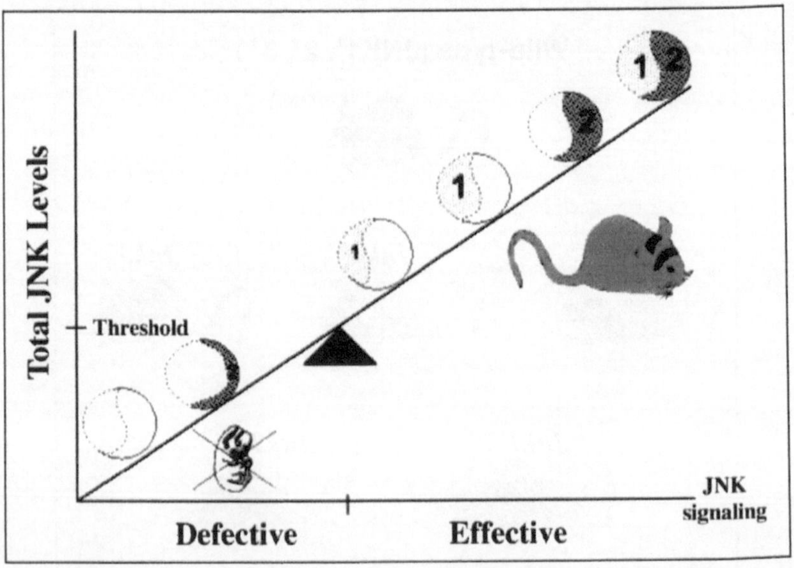

Figure 2. Threshold levels of JNK expression (JNK1 or JNK2) are required for embryonic development. Though *Jnk1*$^{-/-}$ and *Jnk2*$^{-/-}$ mice are viable, absence of both *Jnk* genes result in embryonic lethality. However, gene dosage also appears to play a role in determining the JNK levels in mice. *Jnk1*$^{-/-}$*Jnk2*$^{+/-}$ embryos, but not *Jnk1*$^{+/-}$*Jnk2*$^{-/-}$ embryos, phenocopy the defects found in *Jnk1*$^{-/-}$*Jnk2*$^{-/-}$ embryos. Analysis of total JNK protein levels indicated that the JNK levels were markedly reduced in *Jnk1*$^{-/-}$*Jnk2*$^{+/-}$, but not *Jnk1*$^{+/-}$*Jnk2*$^{-/-}$ embryos, suggesting that JNK1 is a major contributor to JNK protein levels (refer to text for details). Moreover, this indicates that threshold levels of JNK proteins (JNK1 or JNK2) are required for optimal embryonic development, below which defects are manifested. This JNK-dependent embryonic development is attributable to the overlapping functions of JNK1 and JNK2, as JNK3 is not able to compensate for the absence of JNK1 and JNK2.

proliferation and IL-2 production at limiting doses of stimulation.[11,12] Though the effect of combined JNK1 and JNK2 deletion on these processes were not investigated to due lethality of the *Jnk1*$^{-/-}$*Jnk2*$^{-/-}$ mice, some of the phenotypes were also recapitulated in *Jnk1*$^{+/-}$*Jnk2*$^{+/-}$ mice. For instance, defects in anti-CD3 antibody-induced thymocytes apoptosis and mature total T cell proliferation defects were found in *Jnk1*$^{+/-}$*Jnk2*$^{+/-}$ mice, similar to the individual knockouts,[12] suggesting that JNK1 and JNK2 play similar roles in regulating T cell functions. Consistently, JNK activity was compromised in both these mice strains in several cases, being reduced in thymocytes, mature T cells and bronchoalveolar cells.[11,12,22]

Besides these animal experiments, several lines of evidence also suggest that JNK1 and JNK2 play nonredundant roles in JNK signaling in cultured cells in vitro. IL-1-induced AP-1 activity and UV-A-induced Stat3 phosphorylation at serine 727 was compromised in both *Jnk1*$^{-/-}$ and *Jnk2*$^{-/-}$ synoviocytes and fibroblasts, respectively.[23,24] Moreover, JNK-mediated repression TGF-β1 promoter activity was reversed to equal extents by expression of either JNK1 or JNK2.[25] These data together suggest that JNK1 and JNK2 play similar and nonredundant roles in certain circumstance, perhaps equally contributing to JNK activity.

Similar but Nonredundant Roles of JNK2 and JNK3

Similarly, JNK2 and JNK3 have also been shown to play similar and nonredundant roles, especially in the physiology of the brain. In a neurodegeneration mouse model of Parkinson's disease (PD) induced by 1-methyl-4-phenyl-1,2,4,6-tetrahydropyridine, both JNK2 and JNK3, but not JNK1, were found to be required for dopaminergic neuronal cell death.[26] Moreover, both JNK2 and JNK3 were also found to be required for neuronal cell death due to facial and sciatic nerve axotomy in neonates.[27] The dual deletion of JNK2 and JNK3 resulted in a more pronounced effect compared to the single knockouts in both these models, further reiterating the similar but nonredundant roles of JNK2 and JNK3 in neuronal apoptosis.

Specific Effects Caused by JNK1 Deficiency

Extensive analysis has also revealed specific defects due to loss of JNK1 alone. Subsequent to our initial observations that the *Jnk1*$^{-/-}$ mice were smaller compared to their wild-type counterparts, the role of JNK1 was investigated in bone formation and osteoclast differentiation. JNK1, but not JNK2, was found to be specifically activated by the osteoclast-differentiating factor RANKL.[28] Consequently, activation of JNK1, but not JNK2, was found to be required for efficient osteoclastogenesis from bone marrow monocytes (BMMs) and JNK1 specifically protected BMMs from RANKL-induced apoptosis during differentiation. Moreover, consistent with the smaller size of *Jnk1*$^{-/-}$ mice, JNK1, but not JNK2, was specifically found to play a central role in regulating obesity and insulin resistance.[29] JNK1 deficiency was shown to result in decreased obesity and improved insulin sensitivity and enhanced insulin receptor signaling, concomitant to decreased JNK activity. Such effects were not observed with JNK2 mutant mice, where JNK activity was not altered.[29] These results thus provide genetic evidence that activation of JNK1, but not JNK2, was specifically involved in modulating osteoclastogenesis and obesity.

Similarly, several other JNK1-specific effects were found only in lymphocytes from *Jnk1*$^{-/-}$ mice. Purified CD4$^+$ T helper (T_H) cells from *Jnk1*$^{-/-}$ mice hyper-proliferated and were resistant to activation-induced cell death. Moreover, they preferentially differentiated into T_H2 effector cells, indicating that JNK1 plays an inhibitory role during T_H2 differentiation.[14] In vivo infection models using the parasite *Leishmania major*, which is ideal for studying T_H1 induction in the absence of a concomitant induction of T_H2 cells, also supported these in vitro findings. *Jnk1*$^{-/-}$ mice were unable to resolve *Leishmania major* infections and displayed reduced delayed-type hypersensitivity response, which was due to the simultaneous presence of T_H2 cells in these mice, in contrast to their wild-type counterparts.[30] Moreover, in a leukemia mouse model, JNK1 was specifically found to be required for survival and transformation of preB cells, although the effects of JNK2 deficiency was not tested.[31]

Finally, JNK1 was specifically found to be required for maintenance of neuronal microtubules (MTs) by controlling the phosphorylation of microtubule-associated proteins (MAPs). *Jnk1*$^{-/-}$ mice were found to exhibit disrupted anterior commisure tract formation and progressive loss of MTs within axons and dendrites, which was due to hypo-phosphorylation of the MAPs which compromised their ability to bind MTs.[32] JNK2, on the other hand, was not found to be an efficient kinase for MAPs and *Jnk2*$^{-/-}$ mice did not show any of the MT defects. These data together highlight JNK1-specific roles in the adult mice that cannot be replaced by JNK2.

Specific Roles of JNK2

JNK2 specific immune response defects were also noted in mutant mice. Viral infection of cells resulting in activation of the innate antiviral response mediated by type I interferons (IFNs) was impaired in *Jnk2*$^{-/-}$ cells but not in *Jnk1*$^{-/-}$ cells.[33] This was shown to be due to defects in IFN promoter activation that specifically required JNK2.[33] Moreover, loss of JNK2, but not JNK1, resulted in impairment of differentiation of precursor CD4$^+$ cells to T_H1 effector cells in vitro, due to defective IFN-γ production.[13] Consistent with these findings, JNK2-specific effects were manifested in the *ApoE*$^{-/-}$ mouse model of atherosclerosis, the development of which is thought to be dependent on T_H1 effector cells.[34] *ApoE*$^{-/-}$*Jnk2*$^{-/-}$ mice, but not *ApoE*$^{-/-}$*Jnk1*$^{-/-}$ mice, were found to be less prone to atherosclerosis. In addition, JNK2 deficient macrophages were shown to suppress foam cell formation caused by defective uptake and degradation of modified lipoproteins, a process that is crucial for atherogenesis.[34] Consistently, macrophage specific deletion of JNK2 was sufficient to decrease atherogenesis. In the above examples, JNK2-specific JNK activity (but not JNK1-specific activity) was found to be required for the proper functioning of T cells and macrophages,[13,34] indicating a specific role for JNK2 that cannot be compensated for by JNK1.

JNK3-Dependent Physiological Processes

Since JNK3 expression is restricted, most studies have focused on the role of JNK3 in brain development and neuronal physiology. Initial experiments indicated that absence of JNK3 led to reduced excitotoxicity-induced (i.e., kainic-acid) neuronal apoptosis in the hippocampus.[15] This effect was specific to JNK3 and *Jnk1*$^{-/-}$ and *Jnk2*$^{-/-}$ mice did not show similar phenotyopes.[25] Subsequent experiments revealed that nerve-growth factor-deprived sympathetic neurons from JNK3 mutant mice were resistant to cell death compared to their wild-type counterparts.[36] Similarly, cerebral ischemia and hypoxia-induced brain injury was specifically reduced in *Jnk3*$^{-/-}$ mice.[37] Furthermore, hippocampal neurons from *Jnk3*$^{-/-}$ mice were much more resistant to apoptosis in an in vitro model of cerebral ischemia compared to wild-type cells.[37] In the above examples, phosphorylation of c-Jun, which is thought to be critical for JNK-mediated apoptosis, was reduced,[15,36] highlighting the specific and pro-apoptotic role of JNK3 in these processes that cannot be substituted by JNK1 and JNK2. It is also evident from these findings that JNK3 probably contributes to stress-induced JNK signaling and neuronal apoptosis, as in no case was there a difference in basal cell death rates.

Opposite and Antagonistic Functions of JNK1 or JNK2

Besides the redundant or specific roles of the JNKs, genetic analysis using the mutant mice also revealed the existence of distinct and antagonistic roles for JNK1 and JNK2. Analysis of purified immune CD8$^+$ T cells indicated that whereas *Jnk1*$^{-/-}$ cells had a proliferation defect and impaired interleukin-2 (IL-2) receptor expression, *Jnk2*$^{-/-}$ cells hyper-proliferated and produced more IL-2.[38] Consistently, anti-viral immune response against lymphocytic choriomeningitis virus (LCMV) was differently affected in these mice strains. LCMV-infected *Jnk1*$^{-/-}$ mice showed significantly less virus-specific CD8$^+$ cells whereas *Jnk2*$^{-/-}$ mice produced more CD8$^+$ cells.[39]

In line with the proliferation differences noted with CD8$^+$ T cells, other cellular proliferation models also displayed differential effects due to loss of JNK1 or JNK2. Mouse fibroblasts lacking JNK1 was shown to have a proliferation defect whereas *Jnk2*$^{-/-}$ cells displayed a growth advantage due to faster exit of the G1 phase of the cell cycle, compared to their wild-type counterparts.[40] Likewise, *Jnk2*$^{-/-}$ erythroblasts and hepatocytes were also found to exhibit a proliferation advantage.[40,41] Consistently, epidermal proliferation was also differentially regulated by JNK1 and JNK2. Skin epidermis of *Jnk1*$^{-/-}$ mice was thin and the number of proliferating cells of the stratum basale was markedly reduced compared to wild-type mice, whereas *Jnk2*$^{-/-}$ mice displayed keratinocyte hyperplasia, resulting in increased number of epithelial cell layers which was attributed to enhanced proliferation.[20] Furthermore, phorpbol-ester-induced skin tumour formation was found to be significantly increased in *Jnk1*$^{-/-}$ mice and significantly reduced in *Jnk2*$^{-/-}$ mice.[42,43]

Although JNK1 and JNK2 appear to display antagonistic effects on cellular proliferation, JNK activity was differentially regulated only in some cellular systems. CD8$^+$ T cells lacking JNK1 or JNK2 had reduced JNK activity, suggesting that perhaps JNK activity-independent functions of the JNKs may regulate CD8$^+$ T cell proliferation.[38] However, JNK activity correlated with the phenotypes in fibroblasts. Absence of JNK1 resulted in reduced JNK activity, resulting in reduced c-Jun phosophorylation.[40,44,45] By contrast, absence of JNK2 resulted in increased JNK activity, c-Jun phosphorylation and stability. These data indicated that in fibroblasts at least, JNK1 is the major JNK kinase whereas JNK2 was a negative regulator of c-Jun.[40,44] Consistently, only JNK2 was shown to specifically bind to c-Jun and target it for degradation, and absence of JNK2 thus resulted in increased c-Jun stability and activity, thereby probably leading to increased proliferation.[40] Besides fibroblasts, the role of JNK1 and JNK2 on JNK activity has not been carefully analyzed in the other cellular systems.

Nonetheless, consistent with the differential effects on JNK activity, JNK1 and JNK2 mutant fibroblasts were shown to be differentially sensitive to apoptosis. UV- and TNF-α induced apoptosis was compromised in *Jnk1$^{-/-}$* fibroblasts and elevated in *Jnk2$^{-/-}$* cells, correlating with the differential JNK activity in these cells.[44,45] Together, the genetic evidence indicates that JNK1 and JNK2 could play distinct and opposite functions in several cellular systems, thereby differentially regulating proliferation and cell death.

JNK Specific Effects Analyzed with *Jnk1$^{-/-}$Jnk2$^{-/-}$* Cells

Many studies have also attempted to find JNK-specific functions using cells isolated from *Jnk1$^{-/-}$Jnk2$^{-/-}$* embryos. Initial data indicated that *Jnk1$^{-/-}$Jnk2$^{-/-}$* fibroblasts were resistant to UV-induced apoptosis due to a defects in the mitochondrial death signaling pathway, including the failure to release cytochrome c.[46] Moreover, *Jnk1$^{-/-}$Jnk2$^{-/-}$* fibroblasts were also found to be highly resistant to apoptotic effects of IFN.[47] However, *Jnk1$^{-/-}$Jnk2$^{-/-}$* fibroblasts were also found to be sensitive to TNF-mediated apoptosis, leading to the speculation that JNKs provided survival signaling in response to TNF.[48] These data indicate that JNKs can have both apoptotic and survival functions, depending on the stimulating agent.

Besides, one study highlighted that JNKs were required for cellular wound healing, as *Jnk1$^{-/-}$Jnk2$^{-/-}$* fibroblasts were found to be defective in cellular motility.[49] Besides these physiological studies, *Jnk1$^{-/-}$Jnk2$^{-/-}$* fibroblasts have also been used in several biochemical studies that demonstrated a critical role for the JNKs in the expression of type I collagen gene, Bax and others,[50,51] highlighting the essential role of JNKs in gene regulation.

Conclusions

The analysis of the individual or compound JNK mutant mice has revealed the physiological roles performed by the various JNKs. Moreover, controversial issues based on earlier in vitro findings, such as the role of JNKs in apoptosis, have been settled to some extend as the genetic studies have also supported the existence of both apoptotic and anti-apoptotic functions of the JNKs, depending on the cellular and stimulating-signal context. This review has highlighted the specific and nonredundant, as well as redundant roles performed by the different JNKs in various biological processes. The role of JNK3, whose expression is restricted to the brain regions, have been analyzed primarily in neuronal models of injury and apoptosis, as no overt phenotypes have been observed in other parts (Table 1). By contrast, JNK1 and JNK2, which are ubiquitously expressed and share 83% homology between them,[4] appears to have evolved for both redundant and specific roles. Alas, JNK products are highly relevant as has been demonstrated by the crucial roles they play in various physiological processes. There are both cell-type and signal-specific functions that are performed by these proteins, without which, defects are manifested (Table 1). Nonetheless, natural mutations have not been reported in these genes that would affect their crucial functions. This suggests that JNKs play essential and important roles, without which life would be compromised, and also probably explains why JNK1 and JNK2 can at times play redundant but necessary roles. From an evolutionary perspective, the findings obtained up to date with the genetically modified mice highlights the relevance of conserving similar and homologous genes for diverse and varied functions, as well as for "back-up" functions in some instances. These studies would thus serve as a paradigm for other functional studies aiming to elucidate the specific functions of highly similar proteins.

Outlook

With the extensive analysis performed using the JNK mutant mice, several key biological processes controlled by the various JNKs have been identified. Future work would identify other roles for the JNKs and perhaps delve into identifying key process performed by JNK1 and JNK2 together in the adult mice, by the generation of conditional knockout mice strains. Moreover, one issue that has yet to be resolved is if the various isoforms of the *Jnk* genes would be functionally relevant and may perhaps provide further fine refinement of the biological

Table 1. Physiological role of the JNKs as determined from studies using knockout mice

JNK Specificity	Role	Phenotype Identified in Knockout Mice	Refs.
JNK1	Specific (*Jnk1*$^{-/-}$)	Osteoclast differentiation defect	28
		RANKL-mediated apoptosis increased in bone marrow cells	28
		Lean mice due to decreased obesity	29
		Improved insulin signaling	29
		CD4^{+} T cells hyper-proliferated	14
		CD4^{+} T cells were resistant to activation-induced cell death	14
		CD4^{+} T cells preferentially differentiated into Th2 effector cells	14,30
		Lack of survival signals to and transformation of pre-B cells	31
		Defects in neuronal microtubule (MT) maintenance due to reduced phosphorylation of MT-associated proteins	32
JNK2	Specific (*Jnk2*$^{-/-}$)	Impaired innate anti-viral response and IFN activation	33
		Impaired differentiation of CD4+ cells to TH1 effector cells	13
		Reduced atherosclerosis due to suppression of foam cell formation by the macrophages	34
JNK3	Specific (*Jnk3*$^{-/-}$)	Resistance of hippocampal neurons to excitotoxicity-induced apoptosis	15
		Resistance of sympathetic neurons to NGF-deprivation-induced cell death	36
		Reduced brain injury and neuronal apoptosis due to cerebral ischemia and hypoxia	37
JNK1=JNK2	Redundant (*Jnk1*$^{-/-}$ *Jnk2*$^{-/-}$)	Embryonic lethality	17,18
		Defects in embryonic brain development and neural tube closure	17,18
		Defects in optic closure failure and lens development	19
		Epidermis developmental defects in multiple organs	20
JNK1=JNK2	Non-redundant (*Jnk1*$^{-/-}$ or *Jnk2*$^{-/-}$)	Defects in anti-CD3 antibody induced thymocyte apoptosis	11,12
		Reduced concanavalin-A induced hepatocyte apoptosis and liver damage	21
		Increased hyperoxia-induced lung injury and mortality	22
		Defective mature T cell proliferation at limiting stimulation	11,12
		Defective AP-1 activity in IL-1 induced synoviocytes	23
		Reduced UV-A induced Stat3 phosphorylation at serine 727	24
		Repression of TGFb promoter activity	25
JNK2=JNK3	Non-redundant (*Jnk2*$^{-/-}$ or *Jnk3*$^{-/-}$)	Reduced dopaminergic neuronal cell death in mouse Parkinson's disease model	26
		Reduced neuronal cell death due to facial and sciatic nerve axotomy	27
JNK1∞ JNK2	Antagonistic (Opposite effects in *Jnk1*$^{-/-}$ or *Jnk2*$^{-/-}$)	CD8^{+} T cell proliferation and IL-2 signaling is defective in *Jnk1*$^{-/-}$ and augmented in *Jnk2*$^{-/-}$mice	38
		Fibroblast proliferation is defective in *Jnk1*$^{-/-}$ and increased in *Jnk2*$^{-/-}$mice	40
		Keratinocyte proliferation is defective in *Jnk1*$^{-/-}$ and increased in *Jnk2*$^{-/-}$mice	20
		Erythrocyte and hepatocyte proliferation is augmented in *Jnk2*$^{-/-}$but not in *Jnk1*$^{-/-}$ mice	40,41
		TPA-induced skin tumourigenesis in increased in *Jnk1*$^{-/-}$ but decreased in *Jnk2*$^{-/-}$mice	42,43
		UV and TNFa-induced cell death is reduced in *Jnk1*$^{-/-}$ but increased in *Jnk2*$^{-/-}$fibroblasts	44,45

processes controlled by the JNKs. This would require the generation of mice expressing specific isoforms of the *Jnk* genes. Finally, from the application point of view, the genetic studies have also indicated that specific functions regulated by individual JNKs can be of clinical relevance, such as diabetes and neuro-degenerative disease. As such, future work should be focused on generating individual JNK specific inhibitors, which would potentially be useful in specific pathological conditions.

Acknowledgements

We apologize for not being able to cite all the literature due to space constraints. This work was supported by the National Medical Research Council of Singapore (KS).

References

1. Kyriakis JM, Avruch J. pp54 microtubule-associated protein 2 kinase. A novel serine/threonine protein kinase regulated by phosphorylation and stimulated by poly-L-lysine. J Biol Chem 1990; 265(28):17355-63.
2. Hibi M, Lin A, Smeal T et al. Identification of an oncoprotein- and UV-responsive protein kinase that binds and potentiates the c-Jun activation domain. Genes Dev 1993; 7(11):2135-48.
3. Derijard B, Hibi M, Wu IH et al. JNK1: A protein kinase stimulated by UV light and Ha-Ras that binds and phosphorylates the c-Jun activation domain. Cell 1994; 76(6):1025-37.
4. Kallunki T, Su B, Tsigelny I et al. JNK2 contains a specificity-determining region responsible for efficient c-Jun binding and phosphorylation. Genes Dev 1994; 8(24):2996-3007.
5. Mohit AA, Martin JH, Miller CA. p493F12 kinase: A novel MAP kinase expressed in a subset of neurons in the human nervous system. Neuron 1995; 14(1):67-78.
6. Gupta S, Barrett T, Whitmarsh AJ et al. Selective interaction of JNK protein kinase isoforms with transcription factors. EMBO J 1996; 15(11):2760-70.
7. Weston CR, Davis RJ. The JNK signal transduction pathway. Curr Opin Genet Dev 2002; 12(1):14-21.
8. Chen YR, Tan TH. The c-Jun N-terminal kinase pathway and apoptotic signaling. Int J Oncol 2000; 16(4):651-62.
9. Liu ZG, Hsu H, Goeddel DV et al. Dissection of TNF receptor 1 effector functions: JNK activation is not linked to apoptosis while NF-kappaB activation prevents cell death. Cell 1996; 87(3):565-76.
10. Xia Z, Dickens M, Raingeaud J et al. Opposing effects of ERK and JNK-p38 MAP kinases on apoptosis. Science 1995; 270(5240):1326-31.
11. Sabapathy K, Hu Y, Kallunki T et al. JNK2 is required for efficient T-cell activation and apoptosis but not for normal lymphocyte development. Curr Biol 1999; 9(3):116-25.
12. Sabapathy K, Kallunki T, David JP et al. c-Jun NH2-terminal kinase (JNK)1 and JNK2 have similar and stage-dependent roles in regulating T cell apoptosis and proliferation. J Exp Med 2001; 193(3):317-28.
13. Yang DD, Conze D, Whitmarsh AJ et al. Differentiation of CD4+ T cells to Th1 cells requires MAP kinase JNK2. Immunity 1998; 9(4):575-85.
14. Dong C, Yang DD, Wysk M et al. Defective T cell differentiation in the absence of Jnk1. Science 1998; 282(5396):2092-5.
15. Yang DD, Kuan CY, Whitmarsh AJ et al. Absence of excitotoxicity-induced apoptosis in the hippocampus of mice lacking the Jnk3 gene. Nature 1997; 389(6653):865-70.
16. Carboni L, Carletti R, Tacconi S et al. Differential expression of SAPK isoforms in the rat brain. An in situ hybridisation study in the adult rat brain and during post-natal development. Brain Res Mol Brain Res 1998; 60(1):57-68.
17. Sabapathy K, Jochum W, Hochedlinger K et al. Defective neural tube morphogenesis and altered apoptosis in the absence of both JNK1 and JNK2. Mech Dev 1999; 89(1-2):115-24.
18. Kuan CY, Yang DD, Samanta Roy DR et al. The Jnk1 and Jnk2 protein kinases are required for regional specific apoptosis during early brain development. Neuron 1999; 22(4):667-76.
19. Weston CR, Wong A, Hall JP et al. JNK initiates a cytokine cascade that causes Pax2 expression and closure of the optic fissure. Genes Dev 2003; 17(10):1271-80.
20. Weston CR, Wong A, Hall JP et al. The c-Jun NH2-terminal kinase is essential for epidermal growth factor expression during epidermal morphogenesis. Proc Natl Acad Sci USA 2004; 101(39):14114-9.
21. Maeda S, Chang L, Li ZW et al. IKKbeta is required for prevention of apoptosis mediated by cell-bound but not by circulating TNFalpha. Immunity 2003; 19(5):725-37.

22. Morse D, Otterbein LE, Watkins S et al. Deficiency in the c-Jun NH2-terminal kinase signaling pathway confers susceptibility to hyperoxic lung injury in mice. Am J Physiol Lung Cell Mol Physiol 2003; 285(1):L250-7.
23. Han Z, Boyle DL, Chang L et al. c-Jun N-terminal kinase is required for metalloproteinase expression and joint destruction in inflammatory arthritis. J Clin Invest 2001; 108(1):73-81.
24. Zhang Y, Liu G, Dong Z. MSK1 and JNKs mediate phosphorylation of STAT3 in UVA-irradiated mouse epidermal JB6 cells. J Biol Chem 2001; 276(45):42534-42.
25. Ventura JJ, Kennedy NJ, Flavell RA et al. JNK regulates autocrine expression of TGF-beta1. Mol Cell 2004; 15(2):269-78.
26. Hunot S, Vila M, Teismann P et al. JNK-mediated induction of cyclooxygenase 2 is required for neurodegeneration in a mouse model of Parkinson's disease. Proc Natl Acad Sci USA 2004; 101(2):665-70.
27. Keramaris E, Vanderluit JL, Bahadori M et al. c-Jun N-terminal kinase 3 deficiency protects neurons from axotomy-induced death in vivo through mechanisms independent of c-Jun phosphorylation. J Biol Chem 2005; 280(2):1132-41.
28. David JP, Sabapathy K, Hoffmann O et al. JNK1 modulates osteoclastogenesis through both c-Jun phosphorylation-dependent and -independent mechanisms. J Cell Sci 2002; 115(Pt 22):4317-25.
29. Hirosumi J, Tuncman G, Chang L et al. A central role for JNK in obesity and insulin resistance. Nature 2002; 420(6913):333-6.
30. Constant SL, Dong C, Yang DD et al. JNK1 is required for T cell-mediated immunity against Leishmania major infection. J Immunol 2000; 165(5):2671-6.
31. Hess P, Pihan G, Sawyers CL et al. Survival signaling mediated by c-Jun NH(2)-terminal kinase in transformed B lymphoblasts. Nat Genet 2002; 32(1):201-5.
32. Chang L, Jones Y, Ellisman MH et al. JNK1 is required for maintenance of neuronal microtubules and controls phosphorylation of microtubule-associated proteins. Dev Cell 2003; 4(4):521-33.
33. Chu WM, Ostertag D, Li ZW et al. JNK2 and IKKbeta are required for activating the innate response to viral infection. Immunity 1999; 11(6):721-31.
34. Ricci R, Sumara G, Sumara I et al. Requirement of JNK2 for scavenger receptor A-mediated foam cell formation in atherogenesis. Science 2004; 306(5701):1558-61.
35. Brecht S, Kirchhof R, Chromik A et al. Specific pathophysiological functions of JNK isoforms in the brain. Eur J Neurosci 2005; 21(2):363-77.
36. Bruckner SR, Tammariello SP, Kuan CY et al. JNK3 contributes to c-Jun activation and apoptosis but not oxidative stress in nerve growth factor-deprived sympathetic neurons. J Neurochem 2001; 78(2):298-303.
37. Kuan CY, Whitmarsh AJ, Yang DD et al. A critical role of neural-specific JNK3 for ischemic apoptosis. Proc Natl Acad Sci USA 2003; 100(25):15184-9.
38. Conze D, Krahl T, Kennedy N et al. c-Jun NH(2)-terminal kinase (JNK)1 and JNK2 have distinct roles in CD8(+) T cell activation. J Exp Med 2002; 195(7):811-23.
39. Arbour N, Naniche D, Homann D et al. c-Jun NH(2)-terminal kinase (JNK)1 and JNK2 signaling pathways have divergent roles in CD8(+) T cell-mediated antiviral immunity. J Exp Med 2002; 195(7):801-10.
40. Sabapathy K, Hochedlinger K, Nam SY et al. Distinct roles for JNK1 and JNK2 in regulating JNK activity and c-Jun-dependent cell proliferation. Mol Cell 2004; 15(5):713-25.
41. Sabapathy K, Wagner EF. JNK2: A Negative regulator of cellular proliferation. Cell Cycle 2004; 3(12):1520-3.
42. Chen N, Nomura M, She QB et al. Suppression of skin tumorigenesis in c-Jun NH(2)-terminal kinase-2-deficient mice. Cancer Res 2001; 61(10):3908-12.
43. She QB, Chen N, Bode AM et al. Deficiency of c-Jun-NH(2)-terminal kinase-1 in mice enhances skin tumor development by 12-O-tetradecanoylphorbol-13-acetate. Cancer Res 2002; 62(5):1343-8.
44. Hochedlinger K, Wagner EF, Sabapathy K. Differential effects of JNK1 and JNK2 on signal specific induction of apoptosis. Oncogene 2002; 21(15):2441-5.
45. Liu J, Minemoto Y, Lin A. c-Jun N-terminal protein kinase 1 (JNK1), but not JNK2, is essential for tumor necrosis factor alpha-induced c-Jun kinase activation and apoptosis. Mol Cell Biol 2004; 24(24):10844-56.
46. Tournier C, Hess P, Yang DD et al. Requirement of JNK for stress-induced activation of the cytochrome c-mediated death pathway. Science 2000; 288(5467):870-4.
47. Li G, Xiang Y, Sabapathy K et al. An apoptotic signaling pathway in the interferon antiviral response mediated by RNase L and c-Jun NH2-terminal kinase. J Biol Chem 2004; 279(2):1123-31.
48. Lamb JA, Ventura JJ, Hess P et al. JunD mediates survival signaling by the JNK signal transduction pathway. Mol Cell 2003; 11(6):1479-89.

49. Javelaud D, Laboureau J, Gabison E et al. Disruption of basal JNK activity differentially affects key fibroblast functions important for wound healing. J Biol Chem 2003; 278(27):24624-8.
50. Verrecchia F, Wagner EF, Mauviel A. Distinct involvement of the Jun-N-terminal kinase and NF-kappaB pathways in the repression of the human COL1A2 gene by TNF-alpha. EMBO Rep 2002; 3(11):1069-74.
51. Lei K, Nimnual A, Zong WX et al. The Bax subfamily of Bcl2-related proteins is essential for apoptotic signal transduction by c-Jun NH(2)-terminal kinase. Mol Cell Biol 2002; 22(13):4929-42.

Chapter 4

The Biological Function of JNKKs (MKK4/MKK7 Knockout Mice)

Hiroshi Nishina* and Toshiaki Katada

Abstract

Stress-activated protein kinase/c-Jun NH_2-terminal kinase (SAPK/JNK), which belongs to the family of mitogen-activated protein kinase (MAPK), is activated by many types of cellular stresses or extracellular signals. Recent studies, including the analysis with knockout cells and mice, have led to progress towards understanding the molecular mechanism of stress-induced SAPK/JNK activation and the physiological roles of SAPK/JNK in embryonic development and immune responses. Two SAPK/JNK activators, MKK4 and MKK7, are required for full activation of SAPK/JNK, which responds to various stimuli in an all-or-none manner in mouse embryonic stem (ES) cells. SAPK/JNK activation plays essential roles in organogenesis during mouse development by regulating cell proliferation, survival or apoptosis and in immune responses by regulating cytokine gene expression. Furthermore, SAPK/JNK activation is involved in regulation of mRNA stabilization, cell migration, and cytoskeletal integrity. Thus, SAPK/JNK activation by MKK4 and MKK7 has a wide range of functions in mammalian cells.

MAP kinases (MAPKs) are evolutionary conserved signal-transducing enzymes involved in regulation of many cellular events. Several MAPK groups have been identified in mammalian cells, including extracellular signal-regulated kinase (ERK), p38, ERK5, and SAPK/JNK. These MAPKs are activated by their specific MAPK kinases (MAPKKs): ERK by MEK1 (also known as MKK1) and MEK2 (MKK2), p38 by MKK3 and MKK6, ERK5 by MEK5 (MKK5), and SAPK/JNK by MKK4 (SEK1/JNKK1) and MKK7 (SEK2/JNKK2). These MAPKKs are also activated by various MAPKK kinases (MAPKKKs) such as Raf, MLK, MEKK1, TAK1, ASK1, and TPL-2. In this review, we focus on recent progress in the SAPK/JNK group of MAPK-signaling pathways in mouse embryonic stem (ES) cells, embryos, and adult mice by using MKK4 and MKK7 knockout mice (Fig. 1). It is becoming clear that SAPK/JNK activation regulates many cellular processes such as gene expression, cell survival, and apoptosis.[1-2]

Molecular Mechanism of SAPK/JNK Activation by MKK4 and MKK7 in ES Cells

SAPK/JNK is activated not only by many types of cellular stresses including changes in osmolarity, UV irradiation, heat shock, cisplatinum, etoposide, thapsigargin, and tunicamycin, but also by serum, lysophosphatidic acid (LPA), and inflammatory cytokines, interleukin-1β (IL-1β) and tumor necrosis factor TNF-α). The activated SAPK/JNK phosphorylates a number of substrates including transcription factors, c-Jun, Jun D, and ATF-2, to regulate gene

*Corresponding Author: Hiroshi Nishina—Department of Developmental and Regenerative Biology, Medical Research Institute, Tokyo Medical and Dental University, 1-5-45 Yushima, Bunkyo-ku, Tokyo 113-8510, Japan. Email: nishina.dbio@mri.tmd.ac.jp

The JNK Signaling Pathway, edited by Anning Lin. ©2006 Eurekah.com.

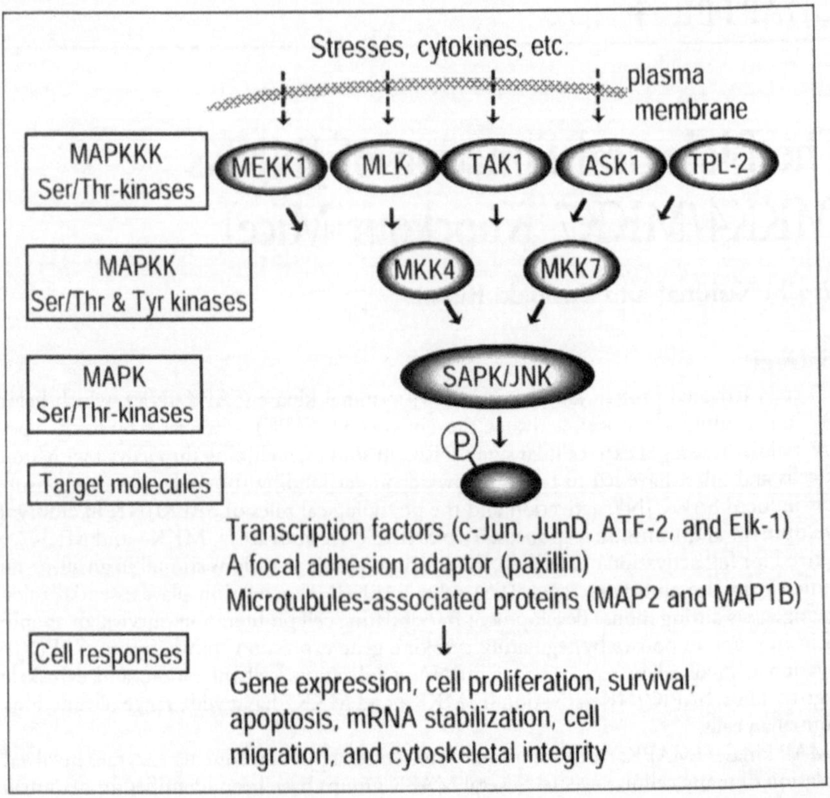

Figure 1. SAPK/JNK-signaling pathways involved in a variety of cell responses. SAPK/JNK is activated by extracellular stimuli including stresses and cytokines through kinase cascades. Activated MAPKKKs such as MEKK1, MLK, TAK1, ASK1, and TPL-2 transmit the signal to two MAPKK, MKK4 and MKK7. SAPK/JNK activated by MKK4 and/or MKK7 phosphorylates target molecules such as c-Jun, paxillin, and MAP2 to regulate a wide range of cell functions.

expression for the stress responses. Activation of SAPK/JNK requires the dual phosphorylation of Tyr and Thr residues located in a Thr-Pro-Tyr motif in the activation loop between VII and VIII of the kinase domain. The phosphorylation is catalyzed by the dual specificity kinases, MKK4 and MKK7, which are capable of catalyzing the phosphorylation of both Thr and Tyr residues.

Recent studies have shown that SEK1 has a preference for the phosphorylation of Tyr residue, while MKK7 for Thr residue of SAPK/JNK in vitro and that both phosphorylation results in the synergistic activation of SAPK/JNK (Fig. 2A).[3] Strong support for this activation mechanism has been obtained from studies of MKK4- and MKK7-gene disruption in ES cells. The severe impairment of SAPK/JNK activation observed in *mkk7*$^{-/-}$ ES cells was accompanied with a loss of the Thr-phosphorylation of SAPK/JNK, without marked reduction in its Tyr-phosphorylated level. On the other hand, Thr-phosphorylation of SAPK/JNK in *mkk4*$^{-/-}$ ES cells was also attenuated, in addition to a decreased level of its Tyr-phosphorylation. These results indicate that the Tyr and Thr residues of SAPK/JNK are sequentially phosphorylated by MKK4 and MKK7, respectively, in stress-stimulated ES cells (Fig. 2B).[4]

Involvement of "scaffold proteins" has been also reported for the correct proceeding of SAPK/JNK-signaling pathway. The scaffold proteins, JIP-1, JIP-2, JIP-3, JSAP1, and JLP, organize the components of SAPK/JNK cascade to secure SAPK/JNK-signaling specificity in

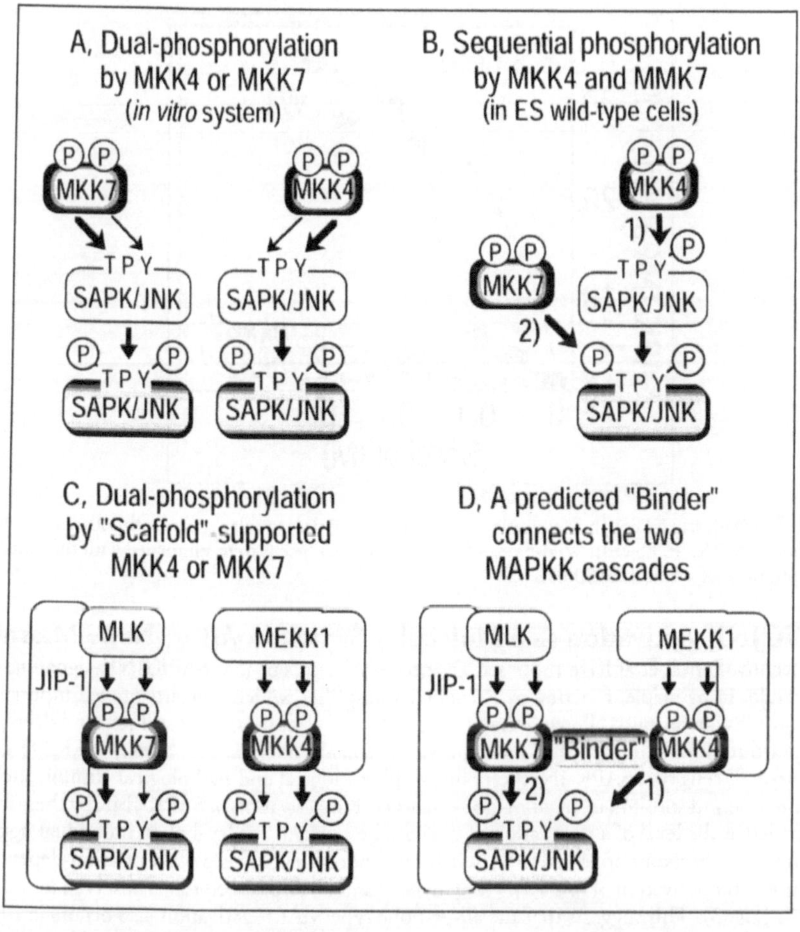

Figure 2. Schematic description of SAPK/JNK phosphorylation by MKK4 and MKK7 under various conditions. A) synergistic activation of SAPK/JNK by the dual-specificity kinase, MKK4 or MKK7, which has been reported in vitro conditions. B) synergistic activation of SAPK/JNK through sequential phosphorylation by MKK4 and MKK7 in murine ES cells. C) activation of SAPK/JNK by MKK4 or MKK7 associated with their scaffold proteins, JIP-1 and MEKK1. D) a predicted "binder" connects the two scaffold-supported MAPKK cascades (MKK4 and MKK7) for the sequential phosphorylation of SAPK/JNK as shown in *panel B. TPY*, Thr-Pro-Tyr motif.

mammalian cells. JIP-1, JIP-2, and JIP-3 bind to SAPK/JNK, MKK7, and mixed-lineage protein kinases (MLKs). JSAP1 is an alternatively spliced variant of JIP3, however, associates with SAPK/JNK, SEK1, and MEKK1. JLP acts as a scaffold protein to bring together Max and c-Myc along with SAPK/JNK and p38, as well as their upstream MKK4 and MEKK3. MEKK1 itself plays a role as a scaffold protein that regulates MKK4 (Fig. 2C). A theoretical consideration indicates that a single set of the scaffold-supported complex, which contains either MKK4 or MKK7, can not induce the sequential phosphorylation, since the synergistic activation of SAPK/JNK require both MKK4 and MKK7 as observed in ES cells. Therefore, two sets of scaffold complexes, one contains MKK4 and the other MKK7, must exist closely in ES cells. Future studies are needed to identify a predicted "binder"-like protein, which connects the two MAPKK (MKK4 and MKK7) cascades (Fig. 2D).

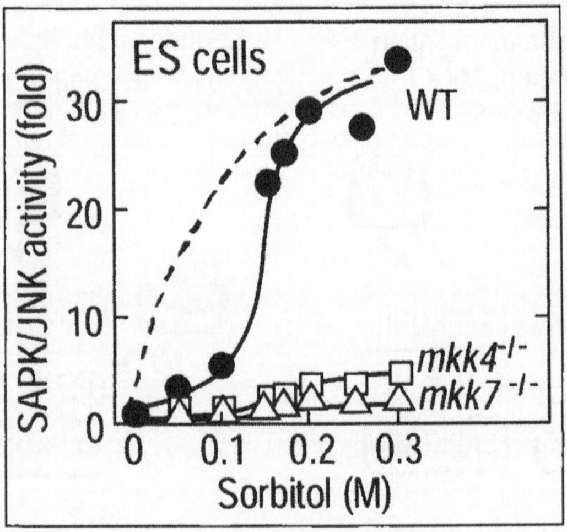

Figure 3. Synergistic SAPK/JNK activation in response to hyper-osmolar stress (sorbitol) requires both MKK4 and MKK7 in ES cells. Wild-type, *sek1*$^{-/-}$, and *mkk7*$^{-/-}$ ES cells were stimulated with the indicated concentrations of sorbitol for 30 min.

SAPK/JNK Activation as a Molecular Switch in All-or-None Manner

Recently, Ferrell et al have proposed an interesting concept that SAPK/JNK-signaling cascade could, in principle, function as a sensitivity amplifier, which converts graded inputs into more switch-like outputs, allowing the cascade to filter out noise and yet still respond decisively to supra-threshold stimuli.[5] They have shown in *Xenopus* oocytes, HeLa cells, HEK293 cells, and Jurkat T cells that SAPK/JNK responds to physiological and pathological stimuli, such as progesterone and sorbitol, in an all-or-none manner. The activation of SAPK/JNK by the stimuli was graded at the level of a population of oocytes, however, at the level of an individual oocyte, the stimulatory response appeared to be switch-like. Indeed, a very steep concentration-dependent response in the activation of SAPK/JNK by hyper-osmolar stress, sorbitol, is observed in murine ES cells (Fig. 3). This suggests that the all-or-none type MAPK activation also occurs in mammalian cells at an individual cell level only when the two MAPKKs are simultaneously activated. Therefore, this MAPK signaling should strictly proceed without errors basically through the two separated signals, one activates MKK4 and another activates MKK7. Although the molecular mechanism whereby the two MAPKKs are simultaneously stimulated by various stress signals remains to be resolved, it is tempting to speculate that the two separated pathways leading to SAPK/JNK activation may exist and physiologically function as a fail-safe mechanism.

Role of SAPK/JNK in Mouse Development

Mkk4, *mkk7*, and all three *Jnk* (*Jnk1, 2*, and *3*) loci have been knocked out at present. *Mkk4*$^{-/-}$ embryos die between embryonic day 10.5 (E10.5) and E12.5 with impaired liver formation and massive apoptosis (Fig. 4). It has been shown that SEK1-mediated SAPK/JNK pathway downstream TNF-α receptor 1 (TNFR1) participates in embryonic hepatoblast proliferation and survival via a pathway different from NF-κB-induced anti-apoptosis.[6] On the other hand, *mkk7*$^{-/-}$ embryos die between E11.5–12.5 with similar defects in liver formation. Impaired proliferation and G2/M cell-cycle kinase CDC2 expression are observed in *mkk7*$^{-/-}$ hepatoblasts.[7] These results indicate that the MKK7-JNK-c-Jun signaling pathway regulates CDC2 expression in hepatoblasts. Thus, SAPK/JNK activation mediated through MKK4 *plus* MKK7 plays indispensable roles in hepatoblast proliferation and survival during mouse embryogenesis (Fig. 5).

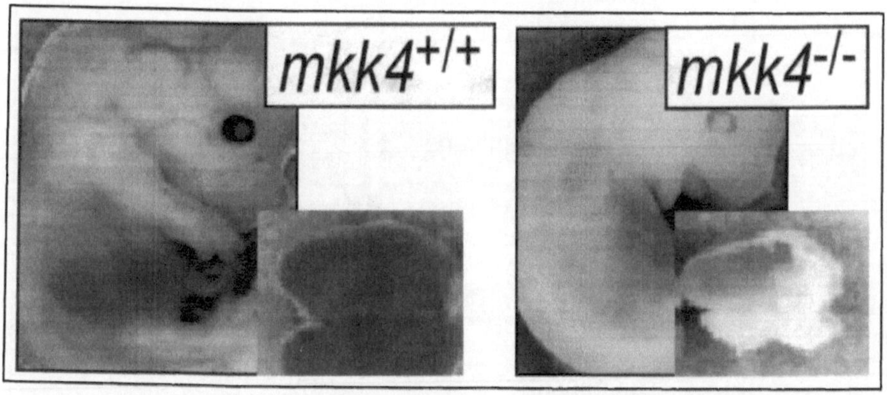

Figure 4. Defective liver formation in *mkk4*$^{-/-}$ embryos. Appearance of wild-type and *mkk4*$^{-/-}$ embryos and livers at embryonic day 12.5. The severe anemia is observed in *mkk4*$^{-/-}$ embryos.

JNK1 and JNK2 are widely expressed in many tissues, but JNK3 is expressed predominantly in nervous system. Mice deficient in the single gene of *Jnk1, Jnk2,* or *Jnk3,* and *Jnk1/Jnk3-* or *Jnk2/Jnk3*-double mutant mice all survived normally. Mice lacking both JNK1 and JNK2 die around E11 with severe dysregulation of apoptosis in brain. Specifically, there was a reduction of cell death in the lateral edges of hindbrain prior to neural tube closure. In contrast, increased apoptosis and caspase activation were found in the mutant forebrain. These results assign both pro-and anti-apoptotic functions to JNK1 and JNK2 in the development of the fetal brain.

Role of SAPK/JNK in Cell Survival and Apoptosis

In mammalian cells, apoptotic signaling cascades can be divided into two broad categories: the intrinsic (mitochondria-dependent) and the extrinsic (death receptor-mediated) pathways. The initiation of mitochondria-dependent pathway requires a change in the organella membrane permeability that is prevented by anti-apoptotic molecules such as Bcl-2 and Bcl-X_L and promoted by pro-apoptotic molecules including Bax and Bak. The permeability change results in the release of mitochondrial proteins. One of the released proteins, cytochrome c, associates with Apaf1 and caspase 9 to activate the effector caspase 3. Cellular stresses such as UV irradiation and heat shock mediate apoptosis through the mitochondria-dependent pathway. However, upstream signaling that regulates the pro-apoptotic molecules remains to be elucidated. Recently, the involvement of SAPK/JNK activation in pro-apoptotic function has been suggested in the study with *Jnk1*$^{-/-}$ *Jnk2*$^{-/-}$ and *mkk4*$^{-/-}$ *mkk7*$^{-/-}$ mouse embryonic fibroblasts (MEFs). Both *Jnk1*$^{-/-}$ *Jnk2*$^{-/-}$ and *mkk4*$^{-/-}$ *mkk7*$^{-/-}$ MEFs exhibited profound defects in stress-induced apoptosis.[8] Furthermore, it has been reported that JNK activation causes the release of apoptogenic factors such as cytochrome c and Smac from isolated mitochondria in a cell-free system. These results strongly suggest that the SAPK/JNK activation directly regulates mitochondria-dependent apoptosis in pro-apoptotic direction.

Recently, we generated ES cells lacking both MKK4 and MKK7 to reevaluate the role of SAPK/JNK activation in the stress-induced and mitochondria-dependent apoptosis.[9] We utilize mouse ES cells in terms of the following advantages: (1) ES cells are a prototype of all cell lineages and can be differentiated into MEF-like cells with retinoic acid, (2) ES cells do not express death receptors including Fas and TNFR1, but have stress-induced, mitochondria-dependent apoptotic pathway, and (3) the molecular mechanism of SAPK/JNK activation is well characterized in ES cells. Interestingly, SAPK/JNK activation is not required for stress-induced and mitochondria-dependent apoptosis in ES and MEF-like cells. Thus, the physiological role of SAPK/JNK activation in cell survival and apoptosis is still

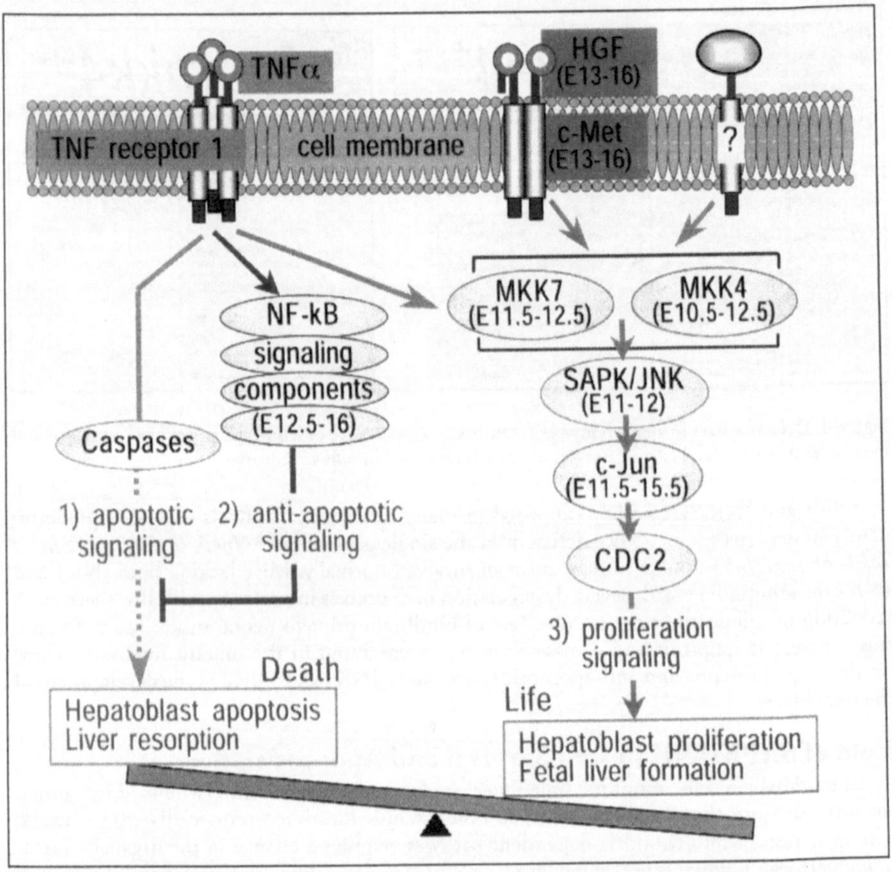

Figure 5. A proposed model for SAPK/JNK signaling pathway in hepatoblasts. The numbers in parentheses are dates of embryonic lethality reported in previous papers. TNFα elicits a wide range of biological responses, such as inflammation, tumor necrosis, differentiation, cell proliferation, and apoptosis, through the stimulation of its receptor, TNFR1. The induction of apoptosis, NF-κB activation, and SAPK/JNK activation are simultaneously mediated through TNFR1. SAPK/JNK activation is involved in cell growth, while activation of NF-κB protects against the apoptosis in hepatoblasts.

controversial, being suggested to have an anti-apoptotic, a pro-apoptotic, or no function in these processes dependent on the types of cells and stimuli.[10]

Role of SAPK/JNK in Mouse Immune Responses

CD4⁺ and C8⁺ T cells are two subsets of peripheral T cells that play important roles during an immune response. After antigen stimulation, CD4⁺ T cells differentiate into effector Th1 or Th2 cells that secret cytokines to help modulate the type of immune response that is generated. Th1 cell promote cell-mediated immunity against intracellular microbial pathogens by expressing interferon-γ, interleukin (IL)-2, and lymphotoxin, whereas Th2 cells promote humoral immunity against parasites and extracellular pathgens by expressing IL-4, IL-5, IL-9, IL-10, and IL-13. CD8⁺ T cells differentiate into cytotoxic T cells to help defend the host during the cell-mediated immune response. Thus, defective T cell development to Th1 and

Th2 cells results in impaired immune responses. Furthermore, inappropriate activation of T cells initiates and perpetuates many autoimmune diseases including rheumatoid, asthma, inflammatory bowel disease, and multiple sclerosis. It has reported that T cells from mice deficient in the *Jnk1* or *Jnk2* gene have a defect in functional differentiation into Th1 or Th2 subsets. One report has shown that JNK1 positively regulates mature T cell activation and that JNK1 and JNK2 have similar and overlapping roles in T cell function. Another report has also reported that JNK1 is required for $CD8^+$ T cell activation, however, that JNK1 and JNK2 have distinct functions in $CD4^+$ T cell differentiation and $CD8^+$ T cell activation. This is analogous to the role of MKK4 in the regulation of T cell activation. One group has shown defective T cell activation, however, another group has shown normal activation of T cells lacking SEK1. These results indicate that the magnitude of the stimuli, the expression of JNK isoforms, and the activation level of SAPK/JNK are important factors determining the direction and efficiency of T cell activation. Thus, SAPK/JNK appears to be involved in the differentiation and activation of T cells, though its precise molecular mechanisms are still controversial.

Other Physiological Roles and Target Molecules of SAPK/JNK

As described above, SAPK/JNK regulates embryonic development including cell proliferation, survival, and apoptosis, and immune responses including T cell differentiation and activation. Furthermore, it has been reported that SAPK/JNK regulates mRNA stabilization, cell migration, and cytoskeletal integrity (Fig. 1). Turnover of mRNA is an important mechanism for the regulation of gene expression in organism from bacteria to mammals. Regulation of mRNA half-life can influence normal cell proliferation, differentiation and oncogenesis. PB-3c mast cells produce IL-3 upon stimulation with extracellular signals, and its short-lived (half-life is about 30 min) mRNA is stabilized by Ca^{2+} ionophores. Using an active MEKK1 and a dominant-negative mutant of JNK, it has been indicated that SAPK/JNK is involved in the regulation of IL-3 mRNA turnover in mast cells. SAPK/JNK is required for *Drosophila* dorsal closure and also essential for cell migration in mammalian cells. Rat bladder tumor epithelial cells (NBT-II) exhibit rapid keratinocyte-like movement. Interestingly, SP600125, a specific inhibitor of JNK, suppresses the movement. Several experiments indicate that JNK1 phosphorylates serine 178 on paxillin, a focal adhesion adaptor, in NBT-II cells. Expression of a paxillin mutant (Ser178 to Ala) inhibited the migration of the cells. Thus, phosphorylation of paxillin by SAPK/JNK seems to be essential for maintaining the labile adhesions required for rapid cell migration. Dynamic assembly and disassembly of microtubules is essential for a variety of cellular functions, such as maintenance of cell morphology and polarity, cell division, cell locomotion, and intracellular trafficking. JNK1-deficient mice exhibit progressive degeneration of long nerve fibers and loss of microtubule integrity in dendrites. Dendritic degeneration of neuronal microtubules is associated with hypo-phosphorylation of microtubule assembly-promoting protein (MAP) 2 and its reduced ability to promote tubulin polymerization. Thus, JNK1 is required for maintaining the cytoskeletal integrity of neuronal cells and is a critical regulator of MAP activity and microtubule assembly.

Conclusions

Data continues to emerge implicating the SAPK/JNK-signaling pathway in a number of physiological functions that may be involved in human disease including autoimmune, anti-inflammatory, neurogenerative diseases, and cancers. In fact, mutations in *Jnk3* gene were identified in human brain tumors. Similarly, mutations in *mkk4* gene have been identified in human cancers of pancreas, lung, breast, colon, and prostate as tumor suppressor genes. Under these circumstances, several pharmaceutical companies have been trying the discovery of SAPK/JNK-related drugs such as an anthrapyrazolone and SP600125. Efforts by many researchers in this field may help to find effective drugs in near future.

References

1. Davis RJ. Signal transduction by the JNK group of MAP kinases. Cell 2000; 103:239-252.
2. Chang L, Karin M. Mammalian MAP kinase signalling cascades. Nature 2001; 410:37-40.
3. Lawler S, Fleming Y, Goedert M et al. Synergistic activation of SAPK1/JNK1 by two MAP kinase kinases in vitro. Curr Biol 1998; 8:1387-1390.
4. Kishimoto H et al. Different properties of SEK1 and MKK7 in dual phosphorylation of stress-induced activated protein kinase SAPK/JNK in embryonic stem cells. J Biol Chem 2003; 278:16595-16601.
5. Bagowski CP, Ferrell JE. Bistability in the JNK cascade. Curr Biol 2001; 11:1176-82.
6. Watanabe T, Nakagawa K, Ohata S et al. SEK1/MKK4-mediated SAPK/JNK signaling participates in embryonic hepatoblast proliferation via a pathway different from NF-κB-induced anti-apoptosis. Dev Biol 2002; 250:332-347.
7. Wada T, Joza N, Cheng HM et al. MKK7 couples stress signalling to G2/M cell-cycle progression and cellular senescence. Nat Cell Biol 2004; 6:215-226.
8. Tournier C, Hess P, Yang DD et al. Requirement of JNK for stress-induced activation of the cytochrome c-mediated death pathway. Science 2000; 288(5467):870-4.
9. Nishitai G, Shimizu N, Negishi T et al. Stress induces mitochondria-mediated apoptosis independent of SAPK/JNK activation in embryonic stem cells. J Biol Chem 2004; 279:1621-1626.
10. Lin A. Activation of the JNK signaling pathway: Breaking the brake on apoptosis. Bio Essays 2002; 25:17-24.

CHAPTER 5

The JNK Signal Transduction Pathway in *Drosophila*

Changwan Lu and Steven X. Hou*

Abstract

The c-Jun N-terminal protein kinases (JNKs) are serine/threonine protein kinases, activated by a wide range of stimuli. Recent studies in genetic systems, especially in *Drosophila*, have shown that this pathway is required in an unusually broad set of biological processes, from embryonic dorsal closure, larval thorax closure, adult wound healing, planar cell polarity, immune response, synaptic plasticity, neuronal cargo transport, to apoptosis and lifespan. Both transcriptional and nontranscriptional responses to the JNK signaling contribute to such versatile roles of this pathway in the model organism.

Introduction

JNKs, also known as stress-activated MAP kinases (SAPKs), are a family of serine/threonine protein kinases in the mitogen-activated protein kinase (MAPK) regime. The proto-oncoprotein c-Jun is the first identified substrate of the JNKs.[1,2] Upon phosphorylation at Ser-63 and Ser-73, c-Jun is translocated into nuclei, forming AP-1 heterodimers with c-Fos to bind DNA and thus regulate gene expression (for reviews, see refs. 3-6).

Certain growth factors, cytokines, heat shock, hyperosmolarity and UV-radiation all activate the JNK signal transduction pathway. The mechanisms utilized by the extracellular stimuli to activate JNKs are currently unknown, although probably involving clustering and internalization of cell-surface receptors.[7] The sequence of known events generally begins with recruitment of adaptor molecules and activation of Ras-family small GTPase (e.g., Ras, Rho, Rac, and Cdc42). The downstream includes three tiers of protein kinases: JNKKKKs (MAPKKKKs), JNKKKs (MAPKKKs) and JNKKs (MAPKKs), forming a chain of sequential phosphorylation. Both JNKKKs and JNKKs are dual-specific kinases, phosphorylating threonines and tyrosines. JNKKs phosphorylate and activate JNKs that directly phosphorylate and activate specific targets such as transcription factors.

The JNK signaling is conserved functionally across phylogenic species, since the *Drosophila* and mammalian counterparts in the pathway can be substituted for each other in functional analyses. As genome-wide mutations have been conventionally generated by P-element insertion, ethyl methanesulfonate and irradiation, *Drosophila*, being an outstanding genetic model, has been used for functional examinations in in vivo contexts. Especially, the embryonic dorsal closure (DC) process has generated the majority of our understanding to JNK signaling.

*Corresponding Author: Steven X. Hou—The Laboratory of Immunobiology, National Cancer Institute at Frederick, Building 560, Room 12-70, Frederick, Maryland 21702-1201, U.S.A. Email: shou@mail.ncifcrf.gov

Structure of the DJNK Pathway

Core Components

The first identified component of the DJNK pathway is *hemipterous* (*hep*),[8] based on a genetic screen for mutations with defects in embryonic DC. The *hep* gene encodes a protein kinase similar to human MKK4/JNKK. Other core components in this pathway are identified thereafter through either a classical genetic screen or a reverse-genetic approach. Mutations of most of the components display similar yet distinct DC defects. Typically, a hole is present at the dorsal side in cuticle preparation as the result of DC defects. These components include: *Drosophila* JNK (DJNK) encoded by *basket* (*bsk*),[9,10] *Drosophila* Jun (DJun),[11-13] *Drosophila* Fos (DFos) encoded by *kayak* (*kay*),[14,15] *Drosophila* JNKKK (DJNKKK) encoded by *slipper* (*slpr*),[16] and *Drosophila* JNKKKK (DJNKKKK) encoded by *misshapen* (*msn*).[17] In the upstream of DJNKKKK/Msn, *Drosophila* Ras-family small GTPases activate the JNK pathway under physiological conditions.[18-21] In the downstream, DJNK/Bsk phosphorylates and activates both DJun and DFos, the transcription factor targets.[22,23]

Downstream Targets

Three downstream targets have been identified, *decapentaplegic* (*dpp*),[13,24] *puckered* (*puc*),[25] and *chickadee* (*chic*).[26,27] Dpp is a member of the transforming growth factor-β (TGF-β) superfamily.[28] Mutations of components in the Dpp signal transduction pathway also cause DC defects, suggesting that the downstream signaling is also required for DC. Puc is a dual-specific MAP kinase phosphatase that may directly downregulate DJNK/Bsk activity through de-phosphorylation, thus forming a negative feedback loop. *Dpp* and *puc* are expressed restrictively at the leading-edge of the advancing epidermis during embryonic DC, in a JNK signaling-dependent manner. The expression of *dpp* and *puc* is thus used as the in vivo reporter for JNK signaling.[11,13] Chic, an actin-binding protein involved in F-actin polymerization, may directly link the DJNK signaling to the cytoskeleton network in the DC process.

Upstream Activators

Based on causing DC defects and regulating the expression of *dpp* and *puc* at LE, upstream activators of the DJNK signaling have been identified. Among these activators are three Src-family members and another novel nonreceptor tyrosine kinase. *Drosophila* has three Src-related kinases, Src42A, Src64, and Tec29. Although a mutation of any one of these kinases alone does not exhibit DC defects, *Src42A Src64* or *Src42A Tec29* double mutant embryos do have DC defects.[29] The F-actin accumulation and expression of *dpp* and *puc* at LE are disrupted in *Src42A Tec29* double mutant embryos. These data suggest that the Src kinases function as upstream activators of the DJNK cascade. Mutations of another nonreceptor tyrosine kinase, Shark, also cause DC defects and disrupt *dpp* expression at LE through regulating the DJNK signaling.[30]

Negative Regulators

In addition to the dual-specific DJNK/Bsk phosphatase Puc, several other proteins have been identified as negative regulators of the DJNK cascade during DC. Mutations of these genes all result in DC defects as well as ectopic overexpression of *dpp* and *puc*. They are *anterior open* (*aop*)/*yan*, *sac1*, *ribbon* (*rib*), *raw*, and *hindsight* (*hnt*). Aop is an ETS-domain protein that is phosphorylated by DJNK/Bsk in vitro. Aop may directly inhibit DJun/DFos-mediated transcription.[13] Sac1 is a lipid phosphatase that dephosphorylates several phosphophatidylinositol phosphates.[31] Mutations of *sac1* disrupt DC and lead to ectopic DJNK activation, suggesting that the phosphoinositide (PI) level is important for DJNK activation during DC. Interestingly, the ectopic *dpp* expression only occurs in stripes as a segmented pattern in *sac1* mutant embryos, although *sac1* mRNA is uniformly expressed in the epidermis, suggesting that the DJNK activation induced by *sac1* mutation is dependent of another protein (or proteins) expressed in a segmented pattern. Other regulators of PI levels such as lipid kinases are likely the candidates. Rib is a protein with a BTB/POZ protein-protein interaction domain.[32] The mechanism of Rib in regulating the DJNK pathway negatively is currently unclear. Raw is a novel protein and HNT is a Zn-finger protein;[33,34] they are predominantly expressed in the amnioserosa during DC. Since *raw* and *hnt* mutations cause ectopic DJNK activation in the epidermis, these two proteins may be involved in the preparation of a ligand in the amnioserosa that eventually inactivates a DJNK component located in the epidermis.

Two PDZ Domain Proteins

Mutations of genes encoding two PDZ domain-containing proteins, Canoe (Cno) and Polychaetoid (Pyd)/ZO-1, also result in DC defects and loss of JNK signaling.[35] Both *dpp* and *puc* expressions are lost in *cno* mutant embryos. *Cno pyd* double mutant embryos show stronger DC defects than that shown by either *cno* or *pyd* single mutant embryos. Cno may function at the upstream of Drac1 because the constitutively activated form of Drac1 is capable of rescuing the *cno* mutant phenotype. The mechanism by which Cno and Pyd activate the DJNK cascade remains unknown.

Scaffold Proteins

A multidomain protein encoded by *connector of kinase to AP-1* (*cka*) may serve as a scaffolding molecule for the DJNK pathway.[23] Embryos lacking the Cka protein show DC defects and reduced *dpp* expression at LE, similar to the consequence of losing JNK pathway components. Cka forms a complex with Hep (DJNKK), Bsk (DJNK), DJun, and DFos. Results from in vivo genetic and cell culture experiments indicate that Cka binding to JNK pathway components contributes to the functioning of the cascade. In cell culture, the complex formation activates Bsk, which in turn phosphorylates and activates DJun and DFos. However, Cka is distinct from scaffolding proteins of the mammalian JNK pathway, in acting as a two-step scaffolding molecule that temporally and spatially organizes the cytosolic and nuclear components of the DJNK pathway, thereby promoting an enhanced activation of these molecules. In cytosol, Cka regulates the DJNK module by bringing Bsk in close proximity to its direct activator Hep. Available data suggest that upon Bsk activation, Cka is phosphorylated to release the Bsk from the complex, which in turn enters into nuclei where it binds and phosphorylates both DJun and DFos. In this aspect, Cka behaves like the yeast scaffolding protein Ste5, which shuttles to the nucleus under certain physiological conditions to prevent kinase activation in the cytosol.[36]

Drosophila Sunday Driver (*syd*) gene encodes a homologue of the mammalian scaffold protein JIP-3.[37] As described below, SYD/DJIP-3 may couple JNK signaling to cargo transportation through assembling a complex of JNK pathway components. However, SYD only functions in neurons and may be a brain-specific scaffolding molecule of the DJNK pathway.

The JNK Pathway Regulates Embryonic Dorsal Closure

After germband retraction in *Drosophila* embryogenesis, an opening of the epidermis is formed at the dorsal side over the amnioserosa. In the subsequent DC process, the epidermis moves across the amnioserosa to close the opening and generates a continuous dorsal epidermis.[38,39] Cell shape changes of the postmitotic epidermis drive DC process.[38] The two symmetric dorsal-most rows of epidermal cells first change shape during the initiation, then the more lateral epidermal cells follow during spreading, and finally the two rows of LE cells fuse at the dorsal midline to finish DC (Fig. 1).[38] The DC process can be divided into three phases: polarization of dorsal-most epidermal (DME) cells (initiation); actin-nucleating centers (ANCs) formation and stabilization; and zippering and sealing.

Polarization of DME Cells

Towards the end of germband retraction, microtubules are still distributed all over the cytoplasm. The initiation of DC is marked by a dorsal-ventral axial alignment of tubulin bundles at

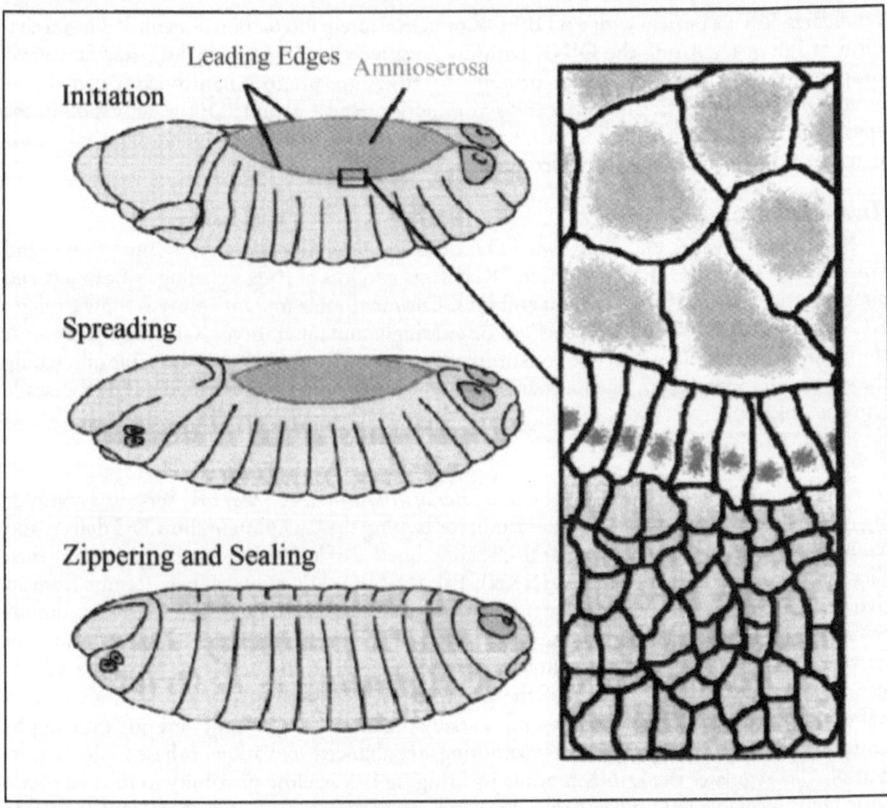

Figure 1. Illustration of the three phases in *Drosophila* embryonic dorsal closure. Initiation begins at stage 13 (11 hr after egg laying). The leading-edge cells (blue) are dorsoventrally elongated. Spreading occurs from stage 13 to stage 15, a 2-hr period. Zippering and sealing is the fusion of the leading-edge cells from opposite sides at the dorsal midline. (Adapted from ref. 38 with copyright permission from Elsevier).

the apical side in DME cells and a dorsal-ventral axial elongation of the DME cells.[40] As the DME cells are polarizing, large quantities of myosin and F-actin appear at the leading edge (LE). The polarized reorganization of the DME cells is reminiscent of events that occur in wing-hair planar polarity formation. Several proteins associated with planar polarity, including Flamingo (Fmi), Frizzled (Fz) and Dishevelled (Dsh), also display polarized distribution in the initiation of DC, whereas preceding the onset of DC, they are spread diffusely in the cytoplasm and irregularly at the cell cortex. In both polarization of the DME cells and the planar polarity formation of wing–hair, the polarization is initiated by an orientation of the microtubules. Mutations in *Wingless* (*wg*) and *dsh* affect the polarization of the DME cells. However, further experiments suggest that Wg signaling has a permissive rather than instructive role.[40] In tissue culture, it has been demonstrated that Cdc42-directed microtubule reorganizations play a key role in cell polarity prior to cell migrations.[41] However, studies in mutations suggest that *cdc42* only affects later phases of DC.[20]

ANCs Formation and Stabilization

The initial accumulation of F-actin occurs at the site of adherens junctions between amnioserosa and DME cells in the form of puncta, which grow over time and extend apically.[40] Eventually, these punta become nucleation centers for the thick accumulation of F-actin and are referred to as ANCs. The ANCs appear to organize the LE cytoskeleton to form filopodia or lamellipodia for later-phase movement of the epidermis.

In *hep* mutants lacking JNK kinase, while the DME cells still polarize and Fmi and Fz also begin to accumulate at the location of ANCs, Fmi and Fz immediately disappear from the feeble ANCs afterwards. JNK signaling appears to affect the maturation and stabilization of the ANCs but not the initial formation of the ANCs. JNK signaling is also required for stimulating the formation of filopodia and lamellipodia at the LE.

Zippering and Sealing

Both phalloidin staining of fixed embryos and imaging of live embryos expressing GFP-tagged actin show that the LE extends filopodia as well as lamellipodia during DC.[39,42] As the LEs of advancing epithelial sheets approach one another, the filopodia and subsequently lamellipodia extending from the LE, seek out, engage, and interdigitate with one another to form weak adhesions (Zippering). These filopodia then regress through actin depolymerization. Finally, the DME cells from opposing side get close enough to form stable adhesion junctions and finish DC (Sealing, Fig. 2).

The Mechanism of JNK Signaling in DC

It appears that the early dorsoventral patterning system must somehow activate the JNK signaling through two Zn-finger transcriptional regulators, *hnt* and *u-shaped* (*ush*).[43] Prior to DC, the JNK signaling is initially active in both amnioserosa and epidermal LE cells (i.e., DME cells during DC process). However, the JNK signaling is down-regulated in the amnioserosa, but not in the LE cells immediately preceding the initiation of DC.[34,44] The JNK activation in LE cells is independent of the *hnt* gene but dependent of the *ush* gene. The JNK signaling is intact in the LE cells in *hnt* mutant embryos, whereas missing in *ush* mutant embryos as indicated by the lack of *dpp* expression.[34,44] The Hnt Zn-finger protein and the JNK phosphatase/Puc are essential for downregulating the JNK signaling in amnioserosa. In both *hnt* and *puc* mutant embryos, the JNK signaling persists in amnioserosa,[34] which interferes with the DC process and results in DC defects. Puc may contribute to the formation of focal complexes/ANCs through downregulating the JNK signaling in the amnioserosa. It has been shown that JNK activation results in nuclear localization of DFos and DJun, while downregulation of the JNK signaling results in relocalization of DFos and DJun to the cytoplasm in the amnioserosa. It appears that formation of ANCs occurs at a boundary of high to low JNK activity that may exist between the LE cells and the amnioserosa. Chic, another downstream target, may also contribute to the formation of ANCs and actin-myosin cables at the LE.

Mutations of genes for the components in the Dpp signal transduction pathway cause DC defects; whereas overexpression of Dpp or a constitutively-activated form of Dpp receptor significantly rescue the DC defects of *Djun* mutant embryos.[11,13] It is without question that the Dpp pathway plays an important role in DC. However, the early events of DC, such as the polarization of DME cells and formation of ANCs, do not require the Dpp signaling.[40] It is proposed that Dpp signaling might downregulate the cytoskeleton at the segment border to release an A-P axial tension that is generated by the epidermal dorsalward movement, therefore allowing a continuous movement of the cell sheet to close the dorsal hole.[45]

Both Wg and Notch (N) pathways crosstalk with the DJNK pathway in DC.[46,47] Embryos deficient in the Wg pathway components show DC defects and reductions in *dpp* and *puc* expression in the LE cells, while ectopic activation of the Wg pathway leads to an expansion of *dpp* expression beyond the LE cells. Wg pathway-driven expression of *dpp* depends on JNK pathway, since a constitutive activation of the Wg pathway can not promote *dpp* expression in *Dfos/kay* mutant embryos.[46] Further, Wg pathway is required for the initial formation of ANCs.[40] Wg pathway may connect to the JNK cascade through the Dsh, as Dsh has been demonstrated to activate the JNK pathway both genetically and biochemically.[48]

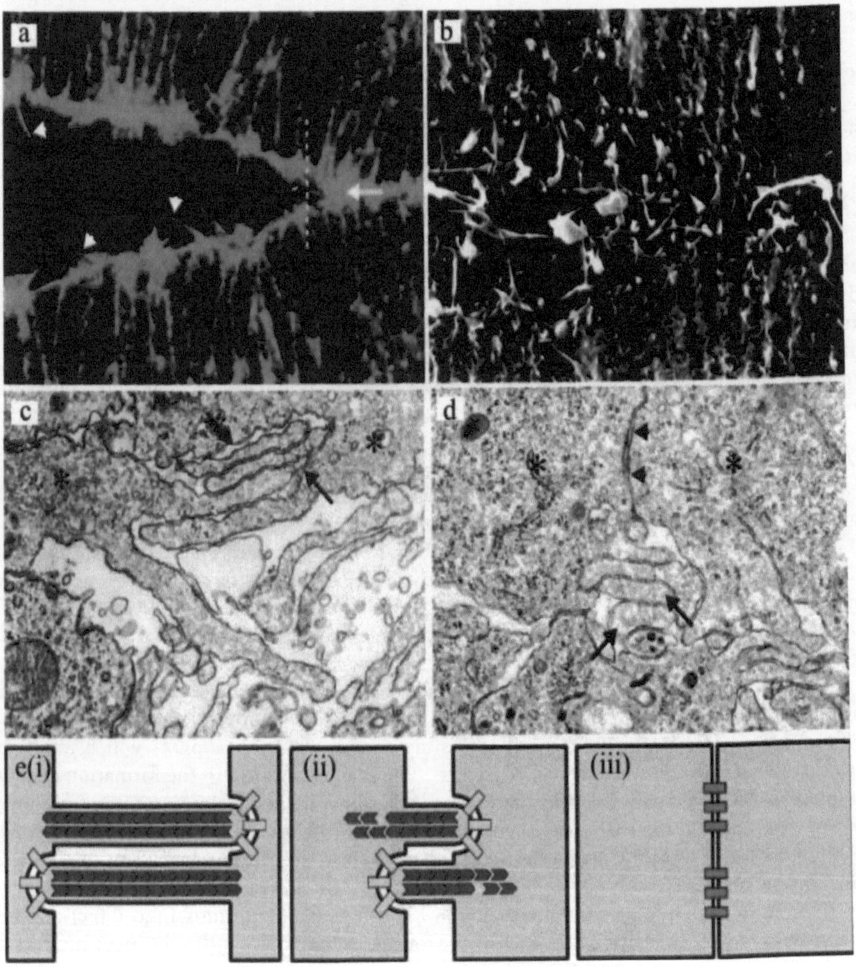

Figure 2. Filopodial function during *Drosophila* dorsal closure. a) Actin-rich filopodia (arrowheads) are extending from epithelial fronts as they zipper up (arrow points to the zippering direction) in GFP-actin expressing embryos. b) Scanning electron microscopy shows zippering protrusions. c) A transmission electron microscopy section shows filopodial interdigitation (arrows) where the leading-edge cells confront one another at the midline (the section plane corresponds to the dashed line in a). d) Further back from the zippering front, filopodia are regressing (arrows) and mature adherens junctions have formed (asterisks). e) Illustration shows three stages of epithelial adhesions evolved from filopodial interactions: i) weak adhesions are formed initially; ii) filopodia regress through actin depolymerization; iii) stable adherens junctions are formed. (Adapted from Martin and Wood, Curr Opin Cell Biol 2002; 14:569-574; with copyright permission from Elsevier.)

In contrast, *N* mutant embryos show DC defects and ectopic expression of *dpp* and *puc* in the epidermis. Embryonic extracts from *N* mutant embryos have a higher JNK activity and a greater ability to phosphorylate Jun than do wild-type extracts.[47] N signaling represses the JNK cascade through a noncanonical pathway, possibly indirectly through the Wg pathway, as N has been shown to inhibit Wg signaling independent of the canonical pathway.[49,50]

Many questions still remain to be answered regarding the DC mechanism. How the three cell types of cells, yolk cells, amnioserosa and DME, communicate and coordinate in the cell sheet movement; how the actin-myosin networks respond to the JNK signaling and organize the driving forces for the movement; how the signaling is coordinated and relayed through the process?

JNK Pathway Regulates Cell Death

JNK Pathway Regulates Apoptosis

In *Drosophila* and mammals, cell death is largely regulated by extracellular ligands, a pathway referred to as 'extrinsic'. The tumor necrosis factor (TNF) is one of the major extracellular ligands that can induce cell death in mammals. *Eiger* is a *Drosophila* homologue of mammalian TNF and Wengen is the Eiger receptor.[51-53] Targeted expression of Eiger in the eyes and wings causes massive cell death and ablation of these organs in *Drosophila.* In Eiger-expressing tissues, the DJNK/Bsk is phosphorylated and expression of *puc*, the downstream target of the JNK pathway, is strongly induced. The cell death and tissue ablation phenotypes can be modified by changing the JNK signaling. Reduction of protein activity by removing one gene copy of either *DTRAF1* (encoding the homolog of human TRAF2),[52,54] *msn*, *hep*, or *bsk* suppresses Eiger-induced apoptosis. Furthermore, *puc* expression is dramatically reduced in *eiger* mutant eye discs, indicating that Eiger functions as a ligand for the DJNK signal transduction pathway under physiological conditions. However, *Djun* is not required for the Eiger-induced cell death, suggesting that *Djun*-mediated new gene transcriptions are not required for the cell death stimulated by the Eiger-JNK pathway.

Different from the mammalian TNF, Eiger does not activate the caspase-8 pathway but rather activates the JNK/caspase-9 pathway.[52] Eiger-activated JNK signaling first upregulates transcription of *hid*, which then activates the *Drosophila* Apaf-1/caspase-9 complex through binding and inactivating the *Drosophila* inhibitor of apoptosis protein 1 (DIAP1). The *Drosophila* data suggest that the ancestral cell death is only regulated by a linear TNF-JNK/apoptosome pathway. At some point during vertebrate evolution, the extrinsic signals (such as TNF) activate caspase-8, while JNKs activate caspase-9 in the intrinsic pathway; the two pathways converge on the effector caspase-3 to execute cell death (Fig. 3).

JNK Pathway Regulates Anoikis

Anoikis, meaning homelessness in ancient Greek, was coined to describe the apoptosis that occurs in cells detached from matrix (or those attached via wrong molecules). JNK signal transduction pathway plays an important role in anoikis.[55] The loss of integrin-mediated cell-matrix contact induces anoikis in certain mammalian cell types.[56] Anoikis is also conserved in *Drosophila.* During embryonic DC, integrins are expressed in the amnioserosa and yolk cell membrane, required for assembly of an intervening extracellular matrix and for the attachment between these two cell layers. Loss of integrin-dependent contact between yolk and amnioserosa results in JNK-dependent programmed anoikis.[57] In wings overexpression of tensin (*blistery*), an adaptor linking integrin to actin filaments, reduces the wing size due to apoptosis through upregulating JNK signaling.[58] During fly wing development, Dpp and Wg signaling cooperatively regulate proximodistal axis formation. Improper levels of Dpp and Wg signaling lead to aberrant morphogenesis in the respective wing zones that thus activates a JNK-dependent apoptosis.[59]

JNK Pathway in Other Processes

JNK pathway is also involved in many other biological processes, although it behaves not exactly the same as in DC. Major evidence is briefly summarized.

Cargo Transportation

Emerging data indicate that defects in axonal transportation contribute to neurodegenerative diseases, such as Alzheimer's disease and Huntington's disease.[60] As the transport tracks, microtubules are polarized with their minus ends aligned towards the cell body and their plus ends

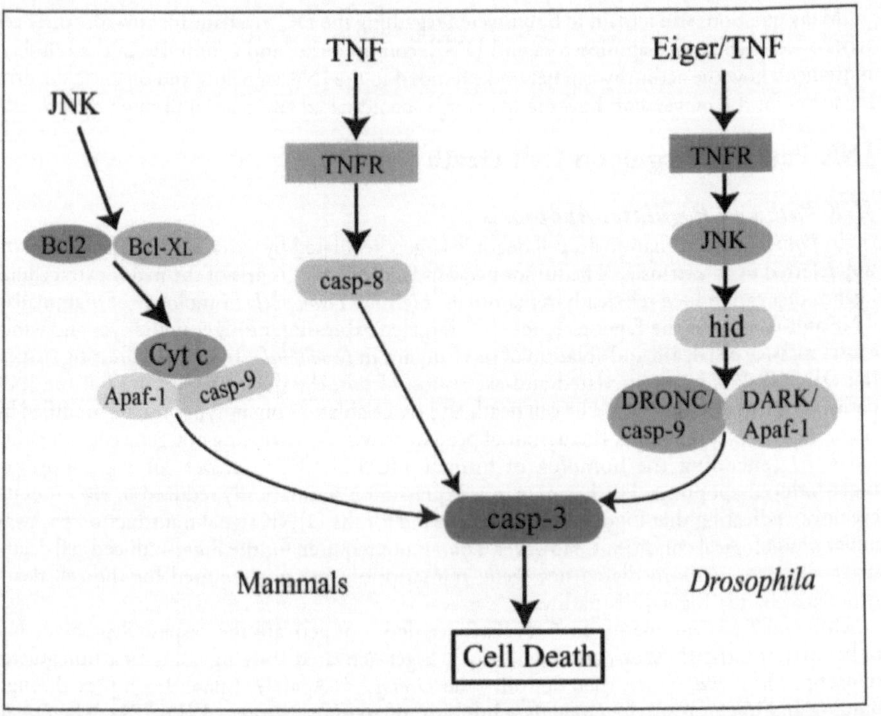

Figure 3. Illustration of JNK-regulated apoptosis in mammals and *Drosophila*. Caspase-3 (casp-3) is the downstream effector in both models. In *Drosophila*, JNK is downstream to TNF. In mammals, TNF and JNK pathways are parallel in the upstream of casp-3. TNF signaling involves caspase-8, while JNK signaling utilizes the cytochrome c (Cyt c) pathway in mitochondria.

towards synaptic terminals. The plus end-directed motor kinesin transports cargos anterogradely from cell body to the nerve terminal, while the minus end-directed motor dynein transports cargos retrogradely. The kinesin and dynein motor proteins are linked to their membrane cargos through scaffolding proteins. It has been reported that the JNK signaling complex forms a 'mobile signalsome' that is actively transported in axons.[60,61]

Three scaffolding proteins for the JNK signaling pathway, the JNK-interacting proteins (JIPs) JIP-1, JIP-2, and JIP-3, are identified as binding partners of the COOH terminus of kinesin light chain in mammals.[61] In *Drosophila*, mutations of *syd* and kinesin-1 cause similar aberrant accumulations of axonal cargos.[37] SYD directly binds the tetratricopeptide repeat domain of kinesin light chain in biochemical experiments. Being the *Drosophila* homologue of mammalian JIP-3, SYD/JIP-3 may represent an adaptor molecule for the transport of a subset of axonal vesicles in brain.

Immune Response

As an invertebrate, *Drosophila* uses innate immunity to oppose microbial invaders. Two major signaling pathways, Toll and Imd, are involved in the immune response.[62] JNK signaling is downstream of Imd, activated by dTAK1. If *Drosophila* SL2 cells are exposed to lipopolysaccharides, the pathogen from gram-negative bacteria, the expression of JNK components and *puc* is upregulated among the immediate early genes, as shown from microarray. In contrast, such upregulation of certain JNK downstream target genes is abolished in *hep* mutation larvae when exposed to septic injury.[63] In addition, *Drosophila* S2 cells exposed to lipopolysaccharides or treated by RNA interference show activation of JNK signaling.[10,64] However, JNK signaling does not induce the production of antimicrobial peptides but rather activates the expression of cytoskeletal proteins in the above models.

Lifespan

Consistent with being a stress responsive signaling, a higher level of JNK signaling extends the lifespan of *Drosophila* adults. *Puc* heterozygotes live about 40% longer than wild-type flies with the same genetic background. The load of oxidized proteins in the *puc* mutants is reduced to about 50%, in line with the free radical theory of aging.[65] Further, *puc* mutants show a higher tolerance to paraquat treatment, an oxidative stress, in a *hep*-dependent manner. Ectopic expression of *hep* in neurons also increases the lifespan significantly, together with a reduction in neuronal sensitivity to paraquat treatment.[65] JNK-regulated neuron death may contribute to fly lifespan. Alternatively, JNK signaling may regulate fly lifespan through the insulin signal transduction pathway. The identification and study of long-lived mutant animals in model organisms, including *Drosophila,* suggest that a revolutionarily conserved, insulin-like signaling network controls aging in all organisms.[66-69]

Oogenesis

In oogenesis, DFos is required for follicular cell migration and reorganization; and the expression of the DFos gene is regulated by *dpp.*[70] The morphogenesis of dorsal appendages and micropyle is also controlled by JNK signaling in oogenesis.[71,72]

Planar Cell Polarity

All cells in the same plane are polarized to the same axis, a structure referred to as planar cell polarity (PCP). PCP is represented in wing hairs, in ommatidia arrangement and in the orientation of 8 photoreceptors in each ommatidium. The Fz/Dsh is the best known signaling responsible for PCP.[73] A hypomorphic *dsh*1 allele causes disorientation of wing hairs and ommatidia;[74] the latter phenotype is partially rescued by ectopic expression of Hep, Bsk or DJun.[75] For a similar phenotype caused by the gain-of-function of Dsh, the ectopic expression of dysfunctional Hep, Bsk or DJun mutations is suppressive, suggesting JNK signaling is activated by Dsh signaling in PCP. Further, Jun phosphorylation is stimulated by Dsh in cell culture assay. In addition, both loss- and gain-of-function of Msn result in defective ommatidial polarity and wing hair formation.[76] However, loss-of-function of Hep or Bsk does not show a phenotype in PCP, suggesting the JNK signaling involved in PCP is redundant to the other MAPK pathways at the downstream of Dsh.[75,77]

Synaptic Plasticity

In response to a long-term stimulation during maturation or learning and memory process, the synaptic activity will be enhanced due to change in gene expression level and modification of focal proteins, a type of synaptic plasticity. Overexpression of both DJun and DFos together in *Drosophila* larval neurons increases the synaptic strength and the bouton number by 30% at the neuromuscular junction, while the expression of dominant-negative DJun and DFos causes the opposite effect.[78] Accordingly, both synaptic strength and bouton number are reduced if Puc or dominant-negative Bsk is ectopically expressed to antagonize the JNK signaling.

Thorax Closure

In metamorphosis from larvae to pupae, epithelial cells in imaginal discs from both sides undergo spreading, evagination and fusion to form a continuous epidermis, a process called thorax closure (TC). Certain *hep* mutant flies are viable but show defects of incomplete metamorphosis at the anterior adult body—typically a clef at the dorsal midline of thoraces.[8,15,79,80] The expression of *puc* was selectively reduced in the peripodial membrane of thoracic discs in *hep* mutants, while ectopic expression of *hep* in the mutant discs induces *puc* expression. *Puc* mutations or *puc* hemizygotes suppress the adult phenotype of *hep* mutants. The mutation of the *Dfos* also causes split thoraxes.[14,80]

Interestingly, in contrast to the regulation by JNK signaling in the LE cells in DC, *dpp* expression is not affected in peripodial cells in *hep* adult mutants.[79] Immunostaining further shows that the phenotype at cellular level is different between *hep* and *dpp* mutations. Larval epidermal cells detach from each other in *hep* mutants, while they remain intact under ectopic expression of a dominant-negative form of Tkv, the Dpp receptor, although imaginal cells spread in neither case.[81] Therefore, JNK and Dpp signaling might be parallel in TC. If an active form of Dcdc42 is expressed in imaginal discs using the UAS-GAL4 system, ectopic *puc* expression and aberrant disc morphology are observed. Thus, small GTPases might also be involved in JNK signaling in metamorphosis.

Wound Healing

In the case of wound healing, epithelia move across the wound area and form a new continuum of epidermis, sharing a high similarity with DC in embryogenesis. In adult flies, *puc* expression is upregulated in epithelia adjacent to the wound site in a DFos-dependent manner, shortly after the wound occurs. Hep phosphorylation is also increased, although *dpp* expression is maintained at the same level.[82] In an embryo model, artificial wounds are created by laser ablation or needle penetration; actin will then accumulates to form a cable structure in LE cells in a RhoA-dependent manner to start the healing process. The wound will not be sealed if Cdc42 function is disrupted,[83,84] similar to the ordeal in DC failure.

Conclusion

The DJNK signal transduction plays an essential role in embryonic DC, in coordinating with Dpp, Wg and N pathways to organize actin-myosin and microtubule networks for step-by-step epithelial movement. The DJNK signaling is not restricted to DC. It has also been demonstrated in many other physiological processes. An interesting question is whether the JNK pathway employs the same cellular artillery in the less understood situations as in DC or uses different machineries. In wound healing, JNK seems to use a DC-similar mechanism, while in aging it may regulate the insulin signal transduction pathway in a way that is different from its classic function in DC.

JNK signaling is self-contained as it induces Puc expression to negatively regulate itself. Since JNK signaling interacts with other signaling, the expression and activity of Puc might also be regulated by such signaling. In morphogenesis such as cell shape change and movement, Dpp is targeted by JNK signaling. In JNK-mediated apoptosis and immune response, the targets are not limited to Dpp. What are the other targets? In the upstream, the linkage between small GTPases and JNK is not addressed yet. Physiological initiators of JNK signaling in the various biological processes are elusive. Finally, microarray assay shows that a wide spectrum of genes are regulated by JNK signaling. Therefore, a persistent pursuit is the only option to reveal more aspects of JNK signaling in development and stress/immune responses.

References

1. Hibi M, Lin A, Smeal T. Identification of an oncoprotein- and UV-responsive protein kinase that binds and potentiates the c-Jun activation domain. Genes Dev 1993; 7:2135-2148.
2. Derijard B, Hibi M, Wu I-H et al. JNK1: A protein kinase stimulated by UV light and Ha-Ras that binds and phosphorylates the c-Jun activation domain. Cell 1994; 76:1025-1037.
3. Davis RJ. Signal transduction by the JNK group of MAP kinases. Cell 2000; 103:239-252.
4. Barr RK, Bogoyevitch MA. The c-Jun N-terminal protein kinase family of mitogen-activated protein kinase (JNK MAPKs). Int J Biochem Cell Biol 2001; 33:1047-1063.
5. Weston CR, Davis RJ. The JNK signal transduction pathway. Curr Opin Genet Dev 2002; 12:14-21.
6. Manning AM, Davis RJ. Targeting JNK for therapeutic benefit: From junk to gold? Nat Rev Drug Dis 2003; 2:554-565.
7. Rosette C, Karin M. Ultraviolet light and osmotic stress: Activation of JNK cascade through multiple growth factor and cytokine receptors. Science 1996; 274:1194-1197.
8. Glise B, Bourbon H, Noselli S. Hemipterous encodes a novel Drosophila MAP kinase kinase, required for epithelial cell sheet movement. Cell 1995; 83:451-461.

9. Riesgo-Escovar JR, Jenni M, Fritz A et al. The Drosophila Jun-N-terminal kinase is required for cell morphogenesis but not for DJUN-dependent cell fate specification in the eye. Genes Dev 1996; 10:2759-2768.
10. Sluss HK, Han Z, Barrett T et al. A JNK signal transduction pathway that mediates morphogenesis and an immune response in Drosophila. Genes Dev 1996; 10:2745-2758.
11. Hou XS, Goldstein ES, Perrimon N. Drosophila Jun relays the Jun amino-terminal kinase signal transduction pathway to the Decapentaplegic signal transduction pathway in regulating epithelial cell sheet movement. Genes Dev 1997; 11:1728-1737.
12. Kockel L, Zeitlinger J, Staszewski LM et al. Jun in Drosophila development: Redundant and nonredundant functions and regulation by two MAPK signal transduction pathways. Genes Dev 1997; 11:1748-1758.
13. Riesgo-Escovar JR, Hafen E. Drosophila Jun kinase regulates expression of decapentaplegic via the ETS-domain protein Aop and the AP-1 transcription factor Djun during dorsal closure. Genes Dev 1997; 11:1717-1727.
14. Riesgo-Escovar JR, Hafen E. Common and distinct roles of DFos and DJun during Drosophila development. Science 1997; 278:669-672.
15. Zeitlinger J, Kockel L, Peverali FA et al. Defective dorsal closure and loss of epidermal decapentaplegic expression in Drosophila fos mutants. EMBO J 1997; 16:7393-7401.
16. Stronach B, Perrimon N. Activation of the JNK pathway during dorsal closure in Drosophila requires the mixed lineage kinase, slipper. Genes Dev 2002; 16:377-387.
17. Su Y-C, Treisman JE, Skolnik EY. The Drosophila Ste20-related kinase misshapen is required for embryonic dorsal closure and acts through a JNK MAPK module on an evolutionarily conserved signaling pathway. Genes Dev 1998; 12:2371-2380.
18. Harden N, Loh HY, Chia W et al. A dominant inhibitory version of the small GTP-binding protein Rac disrupts cytoskeletal structures and inhibits developmental cell shape changes in Drosophila. Development 1995; 121:903-914.
19. Magie CR, Meyer MR, Gorsuch MS et al. Mutations in the Rho1 small GTPase disrupt morphogenesis and segmentation during early Drosophila development. Development 1999; 126:5353-5364.
20. Genova JL, Jong S, Camp JT et al. Functional analysis of Cdc42 in actin filament assembly, epithelial morphogenesis, and cell signaling during Drosophila development. Dev Biol 2000; 221:181-194.
21. Hakeda-Suzuki S, Ng J, Tzu J et al. Rac function and regulation during Drosophila development. Nature 2002; 416:438-442.
22. Bohmann D, Ellis MC, Staszewski LM et al. Drosophila Jun mediates Ras-dependent photoreceptor determination. Cell 1994; 78:973-986.
23. Chen HW, Marinissen MJ, OH SW et al. CKA, a novel multidomain protein, regulates the JUN N-terminal kinase signal transduction pathway in Drosophila. Mol Cell Biol 2002; 22:1792-1803.
24. Sluss HK, Davis RJ. Embryonic morphogenesis signaling pathway mediated by JNK targets the transcription factor Jun and the TGF-beta homologue decapentaplegic. J Cell Biochem 1997; 67:1-12.
25. Martin-Blanco E, Gampel A, Ring J et al. Puckered encodes a phosphatase that mediates a feedback loop regulating JNK activity during dorsal closure in Drosophila. Genes Dev 1998; 12:557-570.
26. Jasper H, Benes V, Schwager C et al. The genomic response of the Drosophila embryo to JNK signaling. Dev Cell 2001; 1:579-586.
27. Cooley L, Verheyen E, Ayers K. Chickadee encodes a profilin required for intercellular cytoplasm transport during Drosophila oogenesis. Cell 1992; 69:173-184.
28. St. Johnston RD, Gelbart WM. Decapentaplegic transcripts are localized along the dorsal-ventral axis of the Drosophila embryo. EMBO J 1987; 6:2785-2791.
29. Tateno M, Nishida Y, Adachi-Yamada T. Regulation of JNK by Src during Drosophila development. Science 2000; 287:324-327.
30. Fernandez R, Takahashi F, Liu Z et al. The Drosophila shark tyrosine kinase is required for embryonic dorsal closure. Genes Dev 2000; 14:604-614.
31. Wei HC, Sanny J, Shu H et al. The Sac1 lipid phosphatase regulates cell shape change and the JNK cascade during dorsal closure in Drosophila. Curr Biol 2003; 13:1882-1887.
32. Bradley PL, Andrew DJ. Ribbon encodes a novel BTB/POZ protein required for directed cell migration in Drosophila melanogaster. Development 2001; 128:3001-3015.
33. Byars CL, Bates KL, Letsou A. The dorsal-open group gene raw is required for restricted DJNK signaling during closure. Development 1999; 126:4913-4923.
34. Reed BH, Wilk R, Lipshitz HD. Downregulation of Jun kinase signaling in the amnioserosa is essential for dorsal closure of the Drosophila embryo. Curr Biol 2001; 11:1098-1108.

35. Miyamoto H, Nihonmatsu I, Kondo S et al. Canoe encodes a novel protein containing a GLGF/DHR motif and functions with Notch and scaborous in common developmental pathways in Drosophila. Genes Dev 1995; 9:612-625.
36. Mahanty SK, Wang Y, Farley FW et al. Nuclear shuttling of yeast scaffold Ste5 is required for its recruitment to the plasma membrane and activation of the mating MAPK cascade. Cell 1999; 98:501-512.
37. Bowman AB, Kamal A, Ritchings BW et al. Kinesin-dependent axonal transport is mediated by the Sunday driver (SYD) protein. Cell 2000; 103:583-594.
38. Noselli S. JNK signaling and morphogenesis in Drosophila. Trends Genet 1998; 14:33-38.
39. Jacinto A, Wood W, Woolner S et al. Dynamic analysis of actin cable function during Drosophila dorsal closure. Curr Biol 2002; 12:1245-1250.
40. Kaltschmidt JA, Lawrence N, Morel V et al. Planar polarity and actin dynamics in the epidermis of Drosophila. Nat Cell Biol 2002; 4:937-944.
41. Etienne-Manneville S, Hall A. Integrin-mediated activation of Cdc42 controls cell polarity in migrating astrocytes through PKCzeta. Cell 2001; 106:489-498.
42. Jacinto A, Wood W, Balayo T et al. Dynamic actin-based epithelial adhesion and cell matching during Drosophila dorsal closure. Curr Biol 2000; 10:1420-1426.
43. Cubadda Y, Heitzler P, Ray RP et al. U-shaped encodes a zinc finger protein that regulates the proneural genes achaete and scute during the formation of bristles in Drosophila. Genes Dev 1997; 11:3083-3095.
44. Stronach BE, Perrimon N. Investigation of leading-edge formation at the interface of amnioserosa and dorsal ectoderm in the Drosophila embryo. Development 2001; 128:2905-2913.
45. Ricos MG, Harden N, Sem KP et al. Dcdc42 acts in TGF-beta signaling during Drosophila morphogenesis: Distinct roles for the Dracl/JNK and Dcdc42/TGF-beta cascades in cytoskeletal regulation. J Cell Sci 1999; 112:1225-1235.
46. McEwen DG, Cox RT, Peifer M. The canonical Wg and JNK signaling cascades collaborate to promote both dorsal closure and ventral patterning. Development 2000; 127:3607-3617.
47. Zecchini V, Brennan K, Martinez-Arias A. An activity of Notch regulates JNK signalling and affects dorsal closure in Drosophila. Curr Biol 1999; 9:460-469.
48. Boutros M, Mlodzik M. Dishevelled: At the crossroads of divergent intracellular signaling pathways. Mech Dev 1999; 83:27-37.
49. Brennan K, Klein T, Wilder E et al. Wingless modulates the effects of dominant negative notch molecules in the developing wing of Drosophila. Dev Biol 1999; 216:210-229.
50. Lawrence N, Langdon T, Brennan K et al. Notch signaling targets the Wingless responsiveness of a Ubx visceral mesoderm enhancer in Drosophila. Curr Biol 2001; 11:375-385.
51. Igaki T, Kanda H, Yamamoto-Goto Y et al. Eiger, a TNF superfamily ligand that triggers the Drosophila JNK pathway. EMBO J 2002; 21:3009-3018.
52. Moreno E, Yan M, Basier K. Evolution of TNF signaling mechanisms: JNK-dependent apoptosis triggered by Eiger, the Drosophila homolog of the TNF superfamily. Curr Biol 2002; 12:1263-1268.
53. Kauppila S, Maaty WSA, Chen P et al. Eiger and its receptor, Wengen, comprise a TNF-like system in Drosophila. Oncogene 2003; 22:4860-4867.
54. Liu H, Su YC, Becker E et al. A Drosophila TNF-receptor-associated factor (TRAF) binds the ste20 kinase Misshapen and activates Jun kinase. Curr Biol 1999; 9:101-104.
55. Cardone MH, Salvesen GS, Widmann C et al. The regulation of anoikis: MEKK-1 activation requires cleavage by caspases. Cell 1997; 90:315-323.
56. Frisch SM, Ruoslahti E. Integrins and anoikis. Curr Opin Cell Biol 1997; 9:701-706.
57. Reed BH, Wilk R, Schock F et al. Integrin-dependent apposition of Drosophila extraembryonic membranes promotes morphogenesis and prevents anoikis. Curr Biol 2004; 14:372-380.
58. Lee SB, Cho KS, Kim E et al. Blistery encodes Drosophila tensin protein and interacts with integrin and the JNK signaling pathway during wing development. Development 2003; 130:4001-4010.
59. Adachi-Yamada T, Fujimura-Kamada K, Nishida Y et al. Distortion of proximodistal information causes JNK-dependent apoptosis in Drosophila wing. Nature 1999; 400:166-169.
60. Gunawardena S, Goldstein LS. Cargo-carrying motor vehicles on the neuronal highway: Transport pathways and neurodegenerative disease. J Neurobiol 2004; 58:258-271.
61. Verhey KJ, Meyer D, Deehan R et al. Cargo of kinesin identified as JIP scaffolding proteins and associated signaling molecules. J Cell Biol 2001; 152:959-970.
62. Hoffmann JA. The immune response of Drosophila. Nature 2003; 426:33-38.
63. Boutros M, Agaisse H, Perrimon N. Sequential activation of signaling pathways during innate immune responses in Drosophila. Dev Cell 2002; 3:711-722.

64. Silverman N, Zhou R, Erlich RL et al. Immune activation of NF-kappaB and JNK requires Drosophila TAK1. J Biol Chem 2003; 278:48928-48934.
65. Wang MC, Bohmann D, Jasper H. JNK signaling confers tolerance to oxidative stress and extends lifespan in Drosophila. Dev Cell 2003; 5:811-816.
66. Guarente L, Kenyon C. Genetic pathways that regulate ageing in model organisms. Nature 2000; 408:255-262.
67. Partridge L, Gems D. Mechanisms of ageing: Public or private? Nat Rev Genet 2002; 3:165-175.
68. Tissenbaum HA, Guarente L. Model organisms as a guide to mammalian aging. Dev Cell 2002; 2:9-19.
69. Tatar M, Bartke A, Antebi A. The endocrine regulation of aging by insulin-like signals. Science 2003; 299:1346-1351.
70. Dequier E, Souid S, Pál M et al. Top-DER and Dpp-dependent requirements for the Drosophila fos/kayak gene in follicular epithelium morphogenesis. Mech Dev 2001; 106:47-60.
71. Suzanne M, Perrimon N, Noselli S. The Drosophila JNK pathway controls the morphogenesis of the egg dorsal appendages and micropyle. Dev Biol 2001; 237:282-294.
72. Dobens LL, Marin-Blanco E, Martinez-Arias A et al. Drosophila puckered regulates Fos/Jun levels during follicle cell morphogenesis. Development 2001; 128:1845-1856.
73. Mlodzik M. Planar cell polarization: Do the same mechanisms regulate Drosophila tissue polarity and vertebrate gastrulation? Trends Genet 2002; 18:564-571.
74. Tree DR, Shulman JM, Rousset R et al. Prickle mediates feedback amplification to generate asymmetric planar cell polarity signaling. Cell 2002; 109:371-381.
75. Boutros M, Paricio N, Strutt DI et al. Dishevelled activates JNK and discriminates between JNK pathways in planar polarity and wingless signaling. Cell 1998; 94:109-118.
76. Paricio N, Feiguin F, Boutros M et al. The Drosophila STE20-like kinase Misshapen is required downstream of the Frizzled receptor in planar polarity signaling. EMBO J 1999; 18:4669-4678.
77. Strutt D, Johnson R, Cooper K et al. Asymmetric localization of frizzled and the determination of notch-dependent cell fate in the Drosophila eye. Curr Biol 2002; 12:813-824.
78. Sanyal S, Sandstrom DJ, Hoeffer CA et al. AP-1 functions upstream of CREB to control synaptic plasticity in Drosophila. Nature 2002; 416:870-874.
79. Agnès F, Suzanne M, Noselli S. The Drosophila JNK pathway controls the morphogenesis of imaginal discs during metamorphosis. Development 1999; 126:5453-5462.
80. Zeitlinger J, Bohmann D. Thorax closure in Drosophila: Involvement of Fos and the JNK pathway. Development 1999; 126:3947-3956.
81. Martin-Blanco E, Pastor-Pareja JC, Garcia-Bellido A. JNK and decapentaplegic signaling control adhesiveness and cytoskeleton dynamics during thorax closure in Drosophila. Proc Natl Acad Sci USA 2000; 97:7888-7893.
82. Rämet M, Lanot R, Zachary D et al. JNK signaling pathway is required for efficient wound healing in Drosophila. Dev Biol 2002; 241:145-156.
83. Wood W, Jacinto A, Grose R et al. Wound healing recapitulates morphogenesis in Drosophila embryos. Nat Cell Biol 2002; 4:907-912.
84. Hutson MS, Tokutake Y, Chang MS et al. Forces for morphogenesis investigated with laser microsurgery and quantitative modeling. Science 2003; 300:145-149.

Chapter 6

Regulation of Apoptosis by the JNK Signaling Pathway

Anning Lin*

Abstract

Regulation of programmed cell death (apoptosis) is one of the biological functions of JNK. Depending on the cell context and the stimulus, JNK can contribute to or suppress apoptosis. In this chapter, I will review the pro- or antiapoptotic functions of JNK in several model systems and discuss the molecular mechanisms underlying the action of JNK in apoptosis and cell survival.

Introduction

c-Jun N-terminal protein kinase (JNK; also known as stress-activated protein kinase, SAPK) is a subfamily of the mitogen-activated protein kinase (MAPK) superfamily.[1] The JNK subfamily has three isoforms (JNK1, 2 and 3), with at least 10 different slicing variants.[2-4] While JNK1 and JNK2 are ubiquitously expressed, JNK3 is mainly expressed in neurons.[2-4] JNK was discovered by virtue of its ability to phosphorylate the transcription factor c-Jun on its N-terminal transactivation domain at two serine residues, Ser63 and Ser73 (see Chapter 1 in this book).[5] The phosphorylation leads to stimulation of c-Jun transcription activity and transformation ability.[1,6,7] Subsequently, it has been reported that JNK also regulates the activity of other transcription factors than c-Jun, including ATF2, Elk-1, p53 and c-Myc[2-4,8] and nontranscription factors, such as members of the Bcl-2 family (Bcl-2, Bim, and BAD).[9-12] JNK is activated by a variety of cellular signals, including proto-oncogenes, growth factors, cytokines, and UV irradiation.[2-4] Like other MAPKs, activation of JNK is mediated by a MAP kinase module, i.e., MAP3K→ MAP2K → MAPK,[5] through sequential protein phosphorylation. Two MAP2Ks (JNKK1/MKK4/SEK1 and JNKK2/MKK7)[13-16] and a group of MAP3Ks (MEKK, ASK1, MLK, TAK1 and TPL-2)[17-21] have been shown to be involved in JNK activation. JNK activity can also be regulated by scaffold proteins such as JIP, β-arrestin and JSAP1, protein phosphatases, and the transcription factor NF-κB.[2-4] It has been implicated that JNK plays a critical role in many physiological and pathological processes in multicellular organisms, such as embryonic development, immune responses, tissue homeostasis, neuronal degeneration, via regulation of cell growth, differentiation, transformation and programmed cell death (apoptosis). This chapter will focus on the role of JNK in cell death and survival in several model systems, with emphasis on the molecular mechanisms by which JNK promotes or suppresses apoptosis.

*Anning Lin—Ben May Institute for Cancer Research, The University of Chicago, 5841 S. Maryland Avenue, MC 6027, Chicago, Illinois 60637, U.S.A. Email: alin@huggins.bsd.uchicago.edu

The JNK Signaling Pathway, edited by Anning Lin. ©2006 Eurekah.com.

JNK and Apoptosis

JNK as a Proapoptotic Protein Kinase

Genetic and biochemical evidence shows that JNK activation is involved in the apoptotic process. Sympathetic neurons isolated from mice deficient in *Jnk*1 and *Jnk*2, or *Jnk*3, were resistant to apoptosis induced by withdrawal of nerve growth factor (NGF) or by the excitotoxic glutamate-receptor agonist kainate treatment.[22,23] Consistently, hippocampal neurons deficient in *Jip*1, one of the putative scaffold proteins that regulate JNK activation, were resistant to kainite-induced apoptosis.[24] In vivo, *Jnk1*$^{-/-}$*Jnk2*$^{-/-}$ mice had defects in neuronal apoptosis during development.[24,25] In addition to neurons, JNK has also been implicated in apoptosis of other cell types. Numerous studies have shown that JNK is activated during apoptosis induced by a variety of death signals, whereas inhibition of JNK activation suppresses or attenuates the apoptosis process.[3,4] It has been shown that in *Jnk*1$^{-/-}$ or *Jnk*1$^{-/-}$*Jnk*2$^{-/-}$ mouse fibroblasts, UV or TNF-α-induced apoptosis was severely impaired.[26,27]

Different JNK isoforms may play distinct roles in apoptosis. While *Jnk1*$^{-/-}$*Jnk2*$^{-/-}$ mice exhibited impaired apoptosis in certain brain regions,[23,25] *Jnk3*$^{-/-}$ mice had normal neuronal development.[28] Interestingly, sympathetic neurons isolated from both mice were resistant to apoptosis induced by NGF withdrawal or kainate treatment.[22] It is possible that JNK1 and JNK2, but not JNK3, are involved in developmental neuronal apoptosis, but all three JNK isoforms are involved in neuronal apoptosis induced by environmental stresses. It has yet to be determined why the neuronal specific JNK3 is not involved in neuronal apoptosis during development.[28] The proapoptotic effect of JNK1/JNK2 in neuronal development is unlikely mediated by c-Jun, since there were no obvious defects in developmental neuronal apoptosis in *c-jun*$^{-/-}$ mice.[29] In fibroblasts, JNK1 and JNK2 play opposite roles in TNF-α or UV-induced apoptosis. TNF-α or UV induced apoptosis was suppressed in *Jnk*1$^{-/-}$ fibroblasts but enhanced in *Jnk*2$^{-/-}$ cells, due to the differential regulation of these two JNK isoforms by TNF-α and UV (see discussion below).[27]

Regulation of the Proapoptotic Activity of JNK

The proapoptotic activity of JNK is regulated by a variety of factors. First of all, the proapoptotic activity of JNK depends on cell types. JNK was proapoptotic in neurons and fibroblasts in response to various death stimuli, such as NGF withdrawal, TNF-α and UV,[22,26,27] but it was antiapoptotic in IL-3-dependent hematopoietic cells.[12] While JNK was required for anti-CD3-induced apoptosis in immature thymocytes, it was not involved in the apoptosis of mature T cells.[25,30] The proapoptotic activity of JNK is also affected by immortalization process. JNK was dispensable for TNF-α-induced apoptosis in primary mouse embryonic fibroblasts (MEFs) in early passages (< 3).[27] In contrast, JNK1 was required for TNF-α-induced apoptosis in later passages (> 8) of primary MEFs or immortalized fibroblasts.[27] These observations may reflect the different mode of TNF-α-induced apoptosis in these cells.[27]

Another major determinant of the proapoptotic activity of JNK is the nature of the death stimulus. MEFs deficient in both *Jnk*1 and *Jnk*2 were resistant to apoptosis induced by UV, anisomycin or MMS, but remained to be sensitive to anti-Fas/CD95 antibody.[26] Immature thymocytes deficient in *Jnk*2 or overexpressing a nonphosphorylatable JNK1(183A/185F) mutant were resistant to anti-CD3 antibody-induced apoptosis, but were still sensitive to anti-Fas/CD95 antibody, UV or dexamethasone-mediated cell death.[31] It appears that only a subset of death signals utilizes JNK to implement apoptosis in a cell type-dependent manner.

The duration of JNK activation also plays a critical role in determining the proapoptotic activity of JNK. Previously, it was observed that prolonged JNK activation was correlated to cell death induced by UV or γ-radiation.[32] In addition, tyrosine phosphatase(s) inhibitors such as vanadate also resulted in prolonged JNK activation and sensitized cells to TNF-α-induced apoptosis.[33] Recently, it was discovered that prolonged JNK activation contributes to TNF-α-induced apoptosis when NF-κB activation is blocked.[3,34,37] Conversion of JNK activation from prolonged to transient suppressed the apoptosis.[35] However, prolonged JNK activation is not sufficient to induce apoptosis (see discussion below).[3,34] The molecular mechanism by which prolonged but not transient JNK activation contributes to apoptosis has yet to be elucidated.

It has long been speculated that the proapoptotic activity of JNK is regulated by other signaling pathways. The best studied case so far is the inhibition by NF-κB on the proapoptotic activity of JNK and TNF-α-induced apoptosis. In fibroblasts deficient in IKKβ or NF-κB/RelA, where NF-κB activation by TNF-α is impaired, TNF-α induced prolonged JNK activation and apoptosis.[34-36] TNF-α also induced prolonged JNK activation and apoptosis in the presence of protein synthesis inhibitors, such as cycloheximide (CHX), which abrogate the activity of the transcription factor NF-κB.[34] These observations demonstrated that in normal cells, in which NF-κB and JNK are simultaneously activated by TNF-α, NF-κB induces expression of inhibitors that suppress JNK activation.[3,37] However, the molecular mechanism underlying the inhibition of JNK by NF-κB is still incompletely understood. Transfection experiments have suggested that X-chromosome-linked IAP (*xiap*) and growth arrest and DNA damage β (*gadd*45β), both of which are NF-κB target genes, may mediate the inhibition of JNK by NF-κB.[3,37] However, TNF-α-induced JNK activation was not affected by inactivation of either gene in mouse fibroblasts (Minemoto Y. and Lin A, unpublished results).[37-39] It is possible that XIAP and Gadd45β may compensate with each other or the true inhibitor(s) is still at large. Recently, it has been shown that NF-κB may induce anti-oxidants to eliminate reactive free radicals (ROS).[40] However, the earliest accumulation of ROS that can be detected takes about ninety minutes while NF-κB-mediated inhibition of JNK occurs as early as thirty minutes after TNF-α stimulation (Kim K, and Lin A, unpublished results).[40,41] Thus, ROS may be only responsible for the late phase (> 90 min) of prolonged JNK activation.

Several components in TNF-α signaling pathway have been proposed to be targeted by NF-κB-mediated inhibition. Although earlier studies did not find JNK itself as a target of NF-κB-mediated inhibition,[34] it has been shown recently that NF-κB inhibits ROS accumulation, which otherwise promotes JNK activation.[42] In the absence of NF-κB activation, TNF-α induces accumulation of ROS, which in turn inactivates several MKPs, resulting in sustained JNK activation and apoptosis.[42] Whether JNK-activating kinase (JNKK1/MKK4 and JNKK2/MKK7) is targeted by NF-κB-mediated inhibition is controversial. Activation of JNKKs (JNKK1/MKK4 and JNKK2/MKK7) by TNF-α was prolonged when NF-κB activation was blocked (Minemoto Y, Kim OK, and Lin A, unpublished results). This suggests that NF-κB-mediated inhibition may also act at or above JNKK. While both JNKKs are common components shared by TNF-α, IL-1 and UV signaling pathways, NF-κB only inhibits JNK activation by TNF-α but not by IL-1[34] or UV (Minemoto Y, Kim OK, and Lin A, unpublished results). Thus, JNKK2/MKK7 is unlikely to be directly targeted by NF-κB-mediated inhibition. It also has been reported that TAK1 is targeted by NF-κB-mediated inhibition in drosophila.[43] However, inhibition of TAK1 also affected activation of another MAPK p38, whose activity is not significantly affected by NF-κB inhibition.[34] It remains to be determined whether TAK1 is negatively regulated by NF-κB in mammalian cells. Taken together, it is likely that NF-κB may inhibit prolonged JNK activation via multiple mechanisms.

Models for Studying the Role of JNK in Apoptosis

JNK Is Essential for UV-Induced Apoptosis

An important model for studying the proapoptotic function of JNK is UV-irradiation-induced apoptosis. UV induces apoptosis via the intrinsic death pathway. UV is probably the strongest activator of JNK, usually inducing prolonged JNK activation.[5] Genetic evidence shows that JNK is essential for UV-induced apoptosis.[26,27] In MEFs deficient in both *Jnk1* and *Jnk2*, UV-induced apoptosis was inhibited.[26] Recently, it has been shown that only JNK1 is required for UV-induced apoptosis in fibroblasts.[27] UV-induced apoptosis was significantly impaired in *Jnk1* null immortalized mouse fibroblasts.[27] In contrast, *Jnk2* null fibroblasts were as sensitive as wildtype cells to UV-induced apoptosis.[27] This is because that UV differentially activates these two JNK isoforms: while JNK1 was significantly activated by UV, activation of JNK2 by UV was greatly reduced.[27] Furthermore, JNK2 even interfered with JNK1 activation by UV in WT cells, probably via competition with JNK1 for upstream activator.[27]

Whether the proapoptotic function of JNK in fibroblasts is mediated by its downstream effector c-Jun in UV-induced apoptosis was controversial. It was reported that c-Jun phosphorylation and activation are required for UV-induced apoptosis. *c-jun*$^{AA63/73}$/*c-jun*$^{AA63/73}$ MEFs, in which the JNK phosphorylation sites Ser63 and Ser73 in c-Jun were replaced by nonphosphorylatable alanines (Ala), were insensitive to UV-induced cell death.[44] This notion was further supported by the finding that in response to UV, c-Jun inhibited p53-mediated cell cycle arrest, thereby promoting p53-mediated apoptosis.[45] In contrast, it was reported that the proapoptotic role of JNK in UV-induced apoptosis was independent of c-Jun-mediated transcription, as UV-induced apoptosis was not affected by the protein synthesis inhibitor cycloheximide or the mRNA synthesis inhibitor actinomycin D.[26] This controversy is likely caused by the different doses of UV irradiation used in these studies.[26,44,45] Recently, we found that c-Jun partially restored the apoptosis response of *Jnk1* null fibroblasts when cells were exposed to low dose of UV but failed to do so when cells were irradiated with high dose of UV.[27] At low dose of UV, p53 is activated and in turn it induces apoptosis when its inhibition on cell cycle is blocked by c-Jun.[45] Under this condition, c-Jun is involved in UV-induced apoptosis.[27,45] However, at high dose, UV does not induce cell cycle arrest, thereby c-Jun could not inhibit the high dose of UV-induced apoptosis.[27] Thus, c-Jun transcription activity is required for low dose of UV-induced apoptosis, but is dispensable for high dose of UV-induced apoptosis.

JNK Contributes to TNF-α-Induced Apoptosis

Another important model for studying the proapoptotic function of JNK is TNF-α-induced apoptosis. As a pleiotropic proinflammatory cytokine, TNF-α activates multiple signaling pathways, such as IκB kinase (IKK), JNK and caspases, thereby it regulates numerous biological activities, including apoptosis.[3,37,46] While caspase activation is required for apoptosis, activation of IKK inhibits apoptosis through the transcription factor NF-κB, whose target genes include members of inhibitors of apoptosis (IAPs) family.[37,47] Therefore, unlike UV, TNF-α only induces cell death when NF-κB activation is inhibited.[37] Recently, it has been discovered that NF-κB-mediated inhibition of JNK activation is also required for suppression of TNF-α-induced apoptosis.[3,27,34-37,42]

TNF-α-induced cell death has characteristics of both apoptosis and necrosis. It is debatable whether JNK contributes to TNF-α-induced apoptosis or necrosis in the absence of NF-κB activation.[34,42,48] Recently, it has been postulated that JNK activation inhibited TNF-α-induced apoptosis but promoted TNF-α-induced necrosis.[48] In contrast, inhibition of JNK by its specific inhibitor SP600125 did not affect TNF-α-induced, reactive oxidant species (ROS)-mediated necrosis when NF-κB activation was blocked.[41] This suggests that JNK may not be required for TNF-α-induced necrosis. Furthermore, in fibroblasts deficient in RelA or IKKβ, TNF-α-induced cell death was significantly blocked by the pan caspase inhibitor z-VAD (Kim K, Tang G, Dibling B and Lin A, unpublished results). This observation shows that JNK contributes to TNF-α-induced apoptosis, rather than necrosis, in the absence of NF-κB activation.

The JNK1 isoform, but not JNK2, contributes to TNF-α-induced apoptosis.[27] Like UV, TNF-α also significantly activates JNK1 but not JNK2.[27] Consequently, TNF-α induced apoptosis was significantly impaired in *Jnk1*-deficient immortalized mouse fibroblasts when NF-κB activation was blocked.[27] In contrast, *Jnk2* null fibroblasts were even more sensitive to TNF-α-induced apoptosis when compared to wildtype cells.[27] This result strongly supports the conclusion that when NF-κB activation is blocked, JNK (JNK1) contributes to TNF-α-induced apoptosis, rather than necrosis.

The Molecular Mechanisms by Which JNK Contributes to Apoptosis

The "Intrinsic Component of Mitochondria-Dependent Death Pathway" Model

Numerous reports have shown that JNK is involved in apoptosis induced by various death stimuli.[2,3,4] *Jnk1*$^{-/-}$ *Jnk2*$^{-/-}$ MEFs were resistant to UV-induced apoptosis, since UV was unable to induce the release of cytochrome C or the depolarization of mitochondrial membrane potential.[26] These observations lead to the hypothesis that JNK is an intrinsic component of the mitochondria-dependent death pathway in apoptosis induced by stresses, such as UV, anisomycin and MMS.[26] Consistent with this hypothesis, it was reported that purified active JNK was able to induce cytochrome c release from purified mitochondria[49] and that microinjection of the constitutively active JNKK2-JNK1 fusion protein in CHO cells was sufficient to induce apoptosis.[50] How activation of JNK induces apoptosis is not clear. It has been reported that the proapoptotic Bcl-2 family protein BAX and BAK are essential for JNK-induced apoptosis since activated JNK was unable to induce apoptosis in fibroblasts deficient in both BAX and BAK.[50] Yet, it has to be determined whether BAX and BAK are indeed the downstream effectors of JNK. Another study has suggested that JNK may induce apoptosis via phosphorylation of the proapoptotic Bcl-2 family protein BIM but the physiological significance of the phosphorylation remains to be determined.[11] In contrast, overexpression of the constitutively active JNKK2-JNK1 fusion protein alone did not induce apoptosis in *RelA*$^{-/-}$ fibroblasts[34] or in other cells including wildtype or *Ikk*β$^{-/-}$ fibroblasts and HeLa cells (Tang G, Kim K, and Lin A, unpublished results). Furthermore, no increased caspase activation or DNA fragmentation occurred in cells that express the constitutively active JNK when compared to cells that express control vector or a kinase-deficient JNKK2(K149M)-JNK1 mutant construct (Tang G, Kim K, and Lin A, unpublished results). Further studies are needed to determine whether JNK activation is sufficient to induce apoptosis or only promotes apoptosis.

The "Breaking the Brake on Apoptosis" Model

Accumulating evidence suggests that JNK may modulate, but not induce, apoptosis. The best studied case so far is the regulation of TNF-α induced apoptosis by JNK activation. In the absence of NF-κB activation, TNF-α induces apoptosis via activation of caspases and prolonged JNK activation.[3,34,35,37,42] Recently it has been shown that JNK1, but not JNK2, is activated by TNF-α and only JNK1 is required for TNF-α induced apoptosis in the absence of NF-κB activation.[27] In contrast, JNK2 activation by TNF-α was negligible and JNK2 even interfered with TNF-α induced JNK1 activation.[27] Consistently, *Jnk2*$^{-/-}$ fibroblasts were more sensitive to TNF-α induced apoptosis than WT fibroblasts.[27] Under certain circumstances, cells undergo apoptosis despite having transient JNK activation or were resistant to apoptosis even having prolonged JNK activation (Liu J, and Lin A, unpublished results). Thus, prolonged JNK activation may only promote apoptosis. It is possible that JNK may function as a regulator to "break the brake" on apoptosis (Fig. 1). This hypothesis is supported by a recent report that prolonged JNK activation resulted in proteolysis of BID.[52] jBID, the novel proteolytic fragment of BID was shown to specifically induce the release of Smac, but not cytochrome c, from mitochondria. Subsequently, Smac displaced caspase8 inhibitor c-IAP1 from TNF-receptor 1 complex, thereby allowing the initiation of apoptosis.[51] However, it has yet to be determined how prolonged, but not transient JNK activation leads to proteolysis of BID and whether jBID indeed mediates the proapoptotic effect of prolonged JNK activation in vivo. Future studies are needed to determine whether the model of "breaking the brake on apoptosis" can explain the proapoptotic role of prolonged JNK activation in apoptosis induced by other death stimuli.

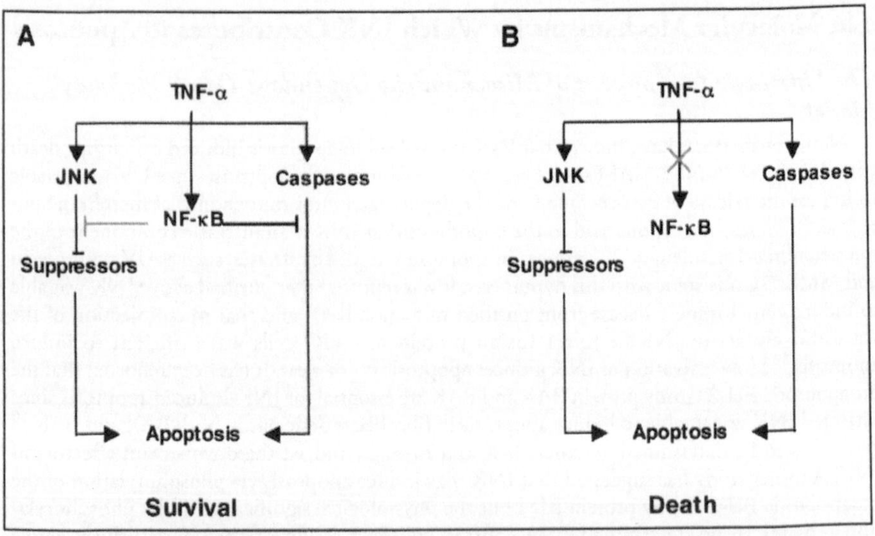

Figure 1. The "breaking the brake" on apoptosis model that explains the role of JNK activation in TNF-α induced apoptosis. A) While caspase activation initiates and executes apoptosis, prolonged JNK activation promotes apoptosis by inactivating suppressors of the mitochondrial-dependent death pathway, i.e., breaking the brake on apoptosis. Activation of NF-κB by TNF-α blocks caspase activation and prevents prolonged JNK activation, thereby inhibiting TNF-α induced apoptosis. B) Inactivation of NF-κB removes inhibition on both caspases and JNK, allowing TNF-α to induce apoptosis.

JNK and Cell Survival

JNK Is Involved in Cell Survival

Despite of its involvement in the process of apoptosis, JNK is not just a proapoptotic protein kinase. Although JNK is activated by many extracellular stimuli, only some of which induce apoptosis. As discussed above, JNK may promote, rather than induce, apoptosis. Evidence has been accumulated that JNK is also involved in cell survival or antiapoptosis. Genetic evidence shows that in *Jnk*1$^{-/-}$*Jnk*2$^{-/-}$ mice, E10.5 hindbrain and forebrain regions had augmented apoptosis.[27] This suggests that JNK may be required for cell survival in these brain regions during development, although the underlying molecular mechanism has yet to be determined. In addition, it has been reported that *Jnkk1*$^{-/-}$ immature thymocytes and peripheral mature T cells were highly sensitive to Fas/CD95 and CD3-mediated apoptosis, suggesting the JNK pathway may be involved in survival of these cells.[31]

The Molecular Mechanism by Which JNK Suppresses Apoptosis

How JNK suppresses apoptosis is incompletely understood. Recent studies show that JNK may suppress apoptosis via phosphorylation and inactivation of the proapoptotic Bcl-2 family protein BAD.[12] In IL-3-dependent hematopoietic cells, in which IL-3 is a survival factor that induces phosphorylation of the pro-apoptotic Bcl-2 family protein BAD. This leads to the sequestration of phosphorylated BAD by the cytoplasmic anchorage protein 14-3-3, thereby preventing inactivation of the pro-survival Bcl-2 family protein Bcl-x_L (Fig. 2).[57] Previously, it has been shown that IL-3-induced phosphorylation of BAD is mediated by Akt and mitochondria-membrane anchored PKA, which phosphorylate BAD at the regulatory serines (Ser112, Ser136 and Ser155).[58] Recently, it has been shown that the survival role of IL-3 is also mediated by JNK.[12] In IL-3 dependent hematopoietic cells, the activity of JNK was decreased upon IL-3 withdrawal but stimulated by IL-3 readdition.[12] Inhibition of JNK enhanced IL-3

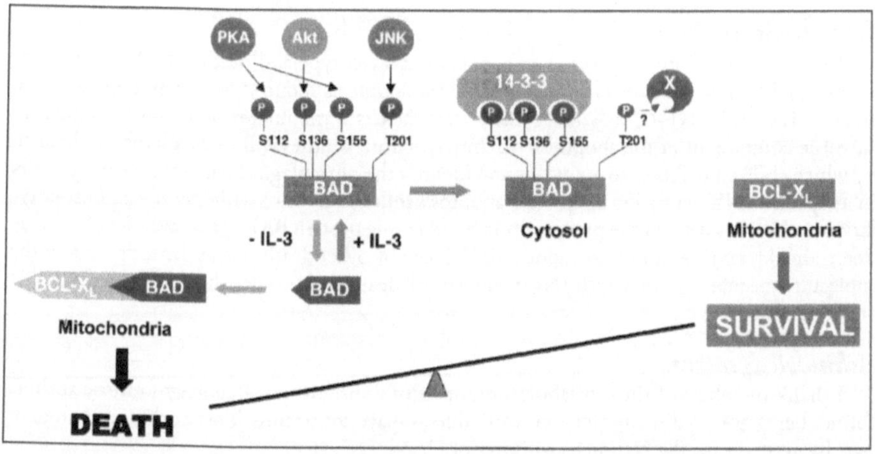

Figure 2. JNK suppresses apoptosis in IL-3-depedent cells via phosphorylation of the pro-apoptotic Bcl-2 family protein BAD. In IL-3-dependent hematopoietic cells, IL-3 activated JNK phosphorylates BAD at Thr201. The phosphorylation results in reduced association of BAD with Bcl-x_L, thereby suppressing apoptosis.

withdrawal-induced apoptosis, whereas a constitutively active JNK suppressed IL-3 withdrawal-induced apoptosis.[12] In contrast to Akt and PKA, who phosphorylate BAD at the regulatory serines, JNK phosphorylated BAD at a novel Thr201 residue.[12] Phosphorylation of BAD at Thr201 by JNK decreased the association between BAD and Bcl-x_L and replacement of Thr201 by the nonphosphorylatable alanine increased the pro-apoptotic activity of BAD.[12] More recently, we found that cotransformation of the constitutively active JNK with BAD significantly reduced its association with Bcl-x_L in yeast (Kim K, and Lin A, unpublished results). In contrast to an early report,[59] JNK does not phosphorylates BAD at Ser128 and phosphorylation of Ser128 has no effects on JNK-dependent apoptosis (Zhang J, Liu J, Yu C, and Lin A, submitted). Taken together, JNK suppresses IL-3 withdrawal-induced apoptosis via phosphorylation and inactivation of the proapoptotic Bcl-2 family protein BAD. The antiapoptotic function of JNK may provide a molecular mechanism by which JNK contributes to B lineage lymphoma/leukemia.[12] Future studies will elucidate how JNK inhibits apoptosis induced by other death stimuli.

JNK Contributes to Tumorigenesis via Regulation of Apoptosis

The pro- or anti-apoptotic functions of JNK can affect tumorigenesis. It has been reported that specific anti-sense oligonucleotides of JNK (JNKAS) suppressed growth of certain types of tumor cells, likely via promotion of apoptosis.[52,53] Interestinly, JNKAS suppressed growth of certain p53-deficient, but not p53 positive, tumor cells.[54] Given the fact that activation of the JNK pathway inhibits p53-induced cell cycle arrest, thereby contributing to p53-induced apoptosis,[36] JNK may only be able to exert its anti-apoptotic function in p53-deficient tumor cells.

Another example is that tumor promoter 12-O-tetradecanolyphorbol-13-acetate (TPA)-induced skin papillomas were suppressed in *Jnk2*$^{-/-}$ mice, due to enhanced apoptosis,[55] while TPA-induced skin tumors were augmented in *Jnk1*$^{-/-}$ mice.[56] However, the underlying molecular mechanism is not clear. Recently, it has been shown that like TNF-α and UV, TPA only significantly activates JNK1, which is involved in apoptosis. In contrast, JNK2 was poorly activated by TPA and it also interfered with JNK activation.[27] Thus, the distinct role of JNK1 and JNK2 on TPA-induced skin papillomas may be due to the differential regulation of JNK1 and JNK2 by TPA.

Conclusions

JNK can contribute to or suppress apoptosis in a cell type and death stimulus-dependent manner. Although JNK activation is required for apoptosis induced by certain death stimuli, such as UV and TNF-α, it is most likely that JNK is a modulator of apoptosis, rather an intrinsic component of the apoptotic machinery. Future studies will elucidate the mechanism by which JNK contributes to apoptosis and identify the physiological relevant target(s) of JNK in apoptosis. JNK activation suppresses apoptosis induced by IL-3 withdrawal via phosphorylation and inactivation of the proapoptotic Bcl-2 family protein BAD. However, it has yet to be determined how JNK suppresses apoptosis induced by other death signals. Investigation of the molecular mechanisms by which JNK regulates cell death and survival should shed light on the JNK biology.

Acknowledgements

I thank members of the Lin laboratory for helpful discussions. I apology to those authors for not being able to directly cite their work due to space constraints. This work is supported in part by grants from the National Institutes of Health (NIH).

References

1. Hibi M, Lin A, Smeal T et al. Identification of an oncoprotein- and UV-responsive protein kinase that binds and potentiates the c-Jun activation domain. Genes Dev 1993; 7:2135-2148.
2. Chang L, Karin M. Mammalian MAP kinase signalling cascades. Nature 2001; 410:37-40.
3. Lin A. Activation of the JNK signaling pathway: Breaking the break on apoptosis. Bioessays 2003; 25:1-8.
4. Davis RJ. Signal transduction by the JNK group of MAP kinases. Cell 2000; 103:239-252.
5. Karin M. The regulation of AP-1 activity by mitogen-activated protein kinases. J Biol Chem 1995; 270:16483-16486.
6. Binetruy B, Smeal T, Karin M. Ha-Ras augments c-Jun activity and stimulates phosphorylation of its activation domain. Nature 1991; 351:122-127.
7. Smeal T, Binetruy B, Mercola DA et al. Oncogenic and transcriptional cooperation with Ha-Ras requires phosphorylation of c-Jun on serines 63 and 73. Nature 1991; 354:494-496.
8. Shaulian E, Karin M. AP-1 as a regulator of cell life and death. Nat Cell Biol 2002; 4:E131-136.
9. Yamamoto K, Ichijo H, Korsmeyer SJ. BCL-2 is phosphorylated and inactivated by an ASK1/Jun N-terminal protein kinase pathway normally activated at G(2)/M. Mol Cell Biol 1999; 19:8469-8478.
10. Maundrell K, Antonsson B, Magnenat E et al. Bcl-2 undergoes phosphorylation by c-Jun N-terminal kinase/stress- activated protein kinases in the presence of the constitutively active GTP-binding protein Rac1. J Biol Chem 1997; 272:25238-25242.
11. Lei K, Davis RJ. JNK phosphorylation of Bim-related members of the Bcl2 family induces Bax-dependent apoptosis. Proc Natl Acad Sci USA 2003; 100:2432-2437.
12. Yu C, Minemoto Y, Zhang J et al. JNK suppresses apoptosis via phosphorylation of the proapoptotic Bcl-2 family protein BAD. Molecular Cell 2004; 13:329-340.
13. Lin A, Minden A, Martinetto H et al. Identification of a dual specificity kinase that activates the Jun kinases and p38-Mpk2. Science 1995; 268:286-290.
14. Derijard B, Raingeaud J, Barrett T et al. Independent human MAP-kinase signal transduction pathways defined by MEK and MKK isoforms. Science 1995; 267:682-685.
15. Sanchez I, Hughes RT, Mayer BJ et al. Role of SAPK/ERK kinase-1 in the stress-activated pathway regulating transcription factor c-Jun. Nature 1994; 372:794-798.
16. Lu X, Nemoto S, Lin A. Identification of c-Jun NH2-terminal protein kinase (JNK)-activating kinase 2 as an activator of JNK but not p38. J Biol Chem 1997; 272:24751-24754.
17. Minden A, Lin A, McMahon M et al. Differential activation of ERK and JNK mitogen-activated protein kinases by Raf-1 and MEKK. Science 1994; 266:1719-1723.
18. Ichijo H, Nishida E, Irie K et al. Induction of apoptosis by ASK1, a mammalian MAPKKK that activates SAPK/JNK and p38 signaling pathways. Science 1997; 275:90-94.
19. Hirai S, Katoh M, Terada M et al. MST/MLK2, a member of the mixed lineage kinase family, directly phosphorylates and activates SEK1, an activator of c-Jun N-terminal kinase/stress-activated protein kinase. J Biol Chem 1997; 272:15167-15173.
20. Yamaguchi K, Shirakabe K, Shibuya H et al. Identification of a member of the MAPKKK family as a potential mediator of TGF-beta signal transduction. Science 1995; 270:2008-2011.

21. Dumitru CD, Ceci JD, Tsatsanis C et al. TNF-alpha induction by LPS is regulated posttranscriptionally via a Tpl2/ERK-dependent pathway. Cell 2000; 103:1071-1083.
22. Bruckner SR, Tammariello SP, Kuan CY et al. JNK3 contributes to c-Jun activation and apoptosis but not oxidative stress in nerve growth factor-deprived sympathetic neurons. J Neurochem 2001; 78:298-303.
23. Kuan CY, Yang D, Samanta Roy DR et al. The Jnk1 and Jnk2 protein kinases are required for regional specific apoptosis during early brain development. Neuron 1999; 22:667-676.
24. Whitmarsh AJ, Kuan CY, Kennedy NJ et al. Requirement of the JIP1 scaffold protein for stress-induced JNK activation. Genes Dev 2001; 15:2421-2432.
25. Sabapathy K, Jochum W, Hochedlinger K et al. Defective neural tube morphogenesis and altered apoptosis in the absence of both JNK1 and JNK2. Mech Dev 1999; 89:115-124.
26. Tournier C, Hess P, Yang DD et al. Requirement of JNK for stress-induced activation of the cytochrome c- mediated death pathway. Science 2000; 288:870-874.
27. Liu J, Minemoto Y, Lin A. JNK1, but not JNK2, is essential for TNF-α induced c-Jun kinase activity and apoptosis. Mol Cell Biol 2004; 24:10844-10856.
28. Yang DD, Kuan CY, Whitmarsh AJ et al. Absence of excitotoxicity-induced apoptosis in the hippocampus of mice lacking the Jnk3 gene. Nature 1997; 389:865-870.
29. Hilberg F, Aguzzi A, Howells N et al. c-jun is essential for normal mouse development and hepatogenesis. Nature 1993; 365:179-181.
30. Sabapathy K, Kallunki T, David JP et al. c-Jun NH2-terminal kinase (JNK)1 and JNK2 have similar and stage-dependent roles in regulating T cell apoptosis and proliferation. J Exp Med 2001; 193:317-328.
31. Nishina H, Fischer KD, Radvanyi L et al. Stress-signalling kinase Sek1 protects thymocytes from apoptosis mediated by CD95 and CD3. Nature 1997; 385:350-353.
32. Chen YR, Wang X, Templeton D et al. The role of c-Jun N-terminal kinase (JNK) in apoptosis induced by ultraviolet C and gamma radiation. Duration of JNK activation may determine cell death and proliferation. J Biol Chem 1996; 271:31929-3136.
33. Guo YL, Baysal K, Kang B et al. Correlation between sustained c-Jun N-terminal protein kinase activation and apoptosis induced by tumor necrosis factor-alpha in rat mesangial cells. J Biol Chem 1998; 273:4027-34.
34. Tang G, Minemoto Y, Dibling B et al. Inhibition of JNK activation through NF-κB target genes. Nature 2001; 414:313-317.
35. Tang F, Tang G, Xiang J et al. Absence of NF-κB-mediated inhibition of c-Jun N-terminal kinase activation contributes to tumor necrosis factor α induced apoptosis. Mol Cell Biol 2002; 22:8571-8579.
36. Maeda S, Chang L, Li Z-W et al. IKKβ is required for prevention of apoptosis mediated by cell-bound but not by circulating TNFα. Immunity 2003; 19:725-737.
37. Karin M, Lin A. NF-κB at the crossroads of life and death. Nat Immunol 2002; 3:221-227.
38. Harlin H, Reffey SB, Duckett CS et al. Characterization of XIAP-deficient mice. Mol Cell Biol 2001; 21:3604-3608.
39. Amanullah A, Azam N, Balliet A et al. Cell survival and a gadd45-factor deficiency. Nature 2003; 424:741-742.
40. Sakon S, Xue X, Takekawa M et al. NF-kappaB inhibits TNF-induced accumulation of ROS that mediate prolonged MAPK activation and necrotic cell death. EMBO J 2003; 22:3898-909.
41. Lin Y, Choksi S, Shen HM et al. Tumor necrosis factor-induced nonapoptotic cell death requires receptor interacting protein-mediated cellular reactive oxygen species accumulation. J Biol Chem 2004; 279:10822-10828.
42. Kamata H, Honda S, Maeda S et al. Reactive oxygen species promote TNFalpha-induced death and sustained JNK activation by inhibiting MAP kinase phosphatases. Cell 2005; 120:649-661.
43. Park JM, Brady H, Ruocco MG et al. Targeting of TAK1 by the NF-kappa B protein Relish regulates the JNK-mediated immune response in Drosophila. Genes Dev 2004; 18:584-594.
44. Behrens A, Sibilia M, Wagner EF. Amino-terminal phosphorylation of c-Jun regulates stress-induced apoptosis and cellular proliferation. Nat Genet 1999; 21:326-329.
45. Shaulian E, Schreiber M, Piu F et al. The mammalian UV response: c-Jun induction is required for exit from p53-imposed growth arrest. Cell 2000; 103:897-907.
46. Baud V, Karin M. Signal transduction by tumor necrosis factor and its relatives. Trends Cell Biol 2001; 11:373-377.
47. Baldwin AS. Control of oncogenesis and cancer therapy resistance by the transcription factor NF-κB. J Clin Invest 2001; 107:241-246.
48. Ventura JJ, Cogswell P, Flavell RA et al. JNK potentiates TNF-stimulated necrosis by increasing the production of cytotoxic reactive oxygen species. Genes Dev 2004; 18:2905-2915.

49. Aoki H, Kang PM, Hampe J et al. Direct activation of mitochondrial apoptosis machinery by c-Jun N-terminal kinase in adult cardiac myocytes. J Biol Chem 2002; 277:10244-10250.
50. Lei K, Nimnual A, Zong WX et al. The Bax subfamily of Bcl-2-related proteins is essential for apoptotic signal transduction by c-Jun NH(2)-terminal kinase. Mol Cell Biol 2002; 22:4929-4942.
51. Deng Y, Ren X, Yang L et al. A JNK-dependent pathway is required for TNF-α induced apoptosis. Cell 2003; 115:61-70.
52. Bost F, McKay R, Bost M et al. The Jun kinase 2 isoform is preferentially required for epidermal growth factor-induced transformation of human A549 lung carcinoma cells. Mol Cell Biol 1999; 19:1938-1949.
53. Potapova O, Anisimov SV, Gorospe M et al. Targets of c-Jun NH(2)-terminal kinase 2-mediated tumor growth regulation revealed by serial analysis of gene expression. Cancer Res 2002; 62:3257-3263.
54. Potapova O, Gorospe M, Dougherty RH et al. Inhibition of c-Jun N-terminal kinase 2 expression suppresses growth and induces apoptosis of human tumor cells in a p53-dependent manner. Mol Cell Biol 2000; 20:1713-1722.
55. Chen N, Nomura M, She QB et al. Suppression of skin tumorigenesis in c-Jun NH(2)-terminal kinase-2- deficient mice. Cancer Res 2001; 61:3908-3912.
56. She QB, Chen N, Bode AM et al. Deficiency of c-Jun-NH(2)-terminal kinase-1 in mice enhances skin tumor development by 12-O-tetradecanoylphorbol-13-acetate. Cancer Res 2002; 62:1343-1348.
57. Zha J, Harada H, Yang E et al. Serine phosphorylation of death agonist BAD in response to survival factor results in binding to 14-3-3 not BCL-X(L). Cell 1996; 87:619-628.
58. Gross A, McDonnell JM, Korsmeyer SJ. BCL-2 family members and the mitochondria in apoptosis. Gene and Dev 1999; 13:1899-1911.
59. Donovan N, Becker EB, Konishi Y et al. JNK phosphorylation and activation of BAD couples the stress-activated signaling pathway to the cell death machinery. J Biol Chem 2002; 277:40944-40949.

Chapter 7

Oxidative Stress, JNK Activation and Cell Death

Han-Ming Shen and Zheng-gang Liu*

Abstract

C-Jun N-terminal kinase (JNK) is one of the three groups of mitogen-activated protein (MAP) kinases that are readily activated by many environmental stimuli including oxidative stress, which refers to the imbalance between enhanced level of reactive oxygen species (ROS) and impaired anti-oxidant defense system. In recent years, an increasing amount of literature has unveiled many important aspects of the regulatory role of ROS in JNK activation. In this Chapter we will first discuss some recent findings regarding the signaling mechanisms of ROS-mediated JNK activation. Second, we will talk about the role of JNK in ROS-mediated cell death (both apoptosis and necrosis) and the diverse signaling pathways involved in the different modes of cell death. Finally, we will analyze the emerging evidence for the involvement of ROS as mediators in the cross-talk between JNK and NF-κB signaling pathway elicited by tumor necrosis factor (TNF). Taken together, advances in understanding the ROS-JNK signaling pathway have shed light on the complex processes deciding the cellular responses to environmental stress.

Introduction

c-Jun N-terminal kinases (JNK) (also known as stress-activated protein kinases) form an important subgroup of the mitogen-activated protein (MAP) kinases superfamily. JNK has three isoforms (JNK1, JNK2 and JNK3) encoded by three genes. The *jnk1* and *jnk2* genes are ubiquitously expressed, while *jnk3* is found to be neural specific.[1] The specific molecular targets of JNK include transcription factor AP-1 (c-Jun, JunB, JunD and ATF-2), p53 and c-Myc,[1,2] as well as many other nontranscription factors such as Bcl-2 family members.[3] The involvement of JNK in controlling diverse cellular functions such as cell proliferation, differentiation, and apoptosis are based on phosphorylation and functional modification of these molecules in stimuli and cell-type dependent manners.

Similar to other members of the MAPK family, JNK activation is mediated by the mammalian MAPK modules comprising MAP kinase kinase kinase (MAPKKK), MAP kinase kinase (MPAKK) and MAPK. JNK is phosphorylated and activated by two MAPKKs (JNNK1/MKK4/SEK1 and JNKK2/MKK7) which possess dual specificity on Thr183 and Tyr185. Upstream of JNKK1 and JNKK2, several MAPKKKs have been identified, including MEK kinase 1 (MEKK1), apoptosis signal-regulating kinase (ASK1), mixed-lineage kinase (MLK), transforming growth factor-beta-associated kinase (TAK1) and TPL2/Cot.[4] Although the physiological relevance of all these MAPKKKs in controlling JNK activation is not completely clear, studies

*Corresponding Author: Zheng-gang Liu—Cell and Cancer Biology Branch, Center for Cancer Research, National Cancer Institute, National Institutes of Health, 9000 Rockville Pike, Bethesda, Maryland, U.S.A. Email: zgliu@helix.nih.gov

The JNK Signaling Pathway, edited by Anning Lin. ©2006 Eurekah.com.

on genetic deletion of these genes have suggested that different stimuli may act through different MAPKKKs to initiate the JNK signaling pathway.[5-7]

JNK are primarily activated by various environmental stresses, including osmotic shock, UV radiation, heat shock, oxidative stress, protein synthesis inhibitors, chemotherapeutic agents, cell death ligand Fas, and proinflammatory cytokines such as tumor necrosis factor-α (TNFα) and interleukin-1 (IL-1). Among them, oxidative stress appears to be particularly important.

Oxidative stress has been defined as an imbalance between the elevated level of reactive oxygen species (ROS) and impaired function of the anti-oxidant defense system.[8] ROS usually include superoxide anion, hydroxyl radicals and hydrogen peroxide (H2O2) that are capable of reacting with and damaging various molecular targets including DNA, protein and lipid. More importantly, ROS have been found to act as signaling molecules involving in various cellular signaling pathways as a model of redox regulation.[9,10] In this chapter, we aim to discuss recent findings on the regulatory role of ROS in JNK activation, especially on the following aspects: (i) the molecular mechanisms of ROS-mediated JNK activation process, (ii) the role of JNK in oxidative stress-mediated cell death (both apoptosis and necrosis), and (iii) the involvement of ROS in the cross-talk between JNK and NF-κB in TNF-induced cell death.

ROS-Mediated JNK Activation

JNK Activation Induced by Both Exogenous and Endogenous ROS

It is known that ROS are potent inducers of JNK. Most reports on ROS-induced JNK activation are resulted from using exogenous ROS, mostly H2O2.[11] Although H2O2 is a relatively weak oxidant comparing to other ROS such as hydroxyl radicals, it has emerged as a particularly important signaling molecule based on its unique biochemical properties: H2O2 is ubiquitously present in the biological system with a relatively long half-life; and more importantly H2O2 is soluble in both lipid and aqueous environments, and thus, capable of reaching its cellular targets when applied extracellularly.[12,13] In addition, JNK activation was also observed in cells treated with some direct ROS precursors such as menadione which gives rise to superoxide radical upon intracellular metabolism.[14,15] However, perhaps biologically more important are endogenously generated ROS that also readily activate JNK, a process often to be critical in determining the fate of the cell.[10,16] Enhanced ROS production is frequently observed in cells exposed to many stimuli such as UV,[17,18] ionizing radiation,[19] cancer chemotherapeutics,[20,21] and TNFα.[22] The main source of intracellularly generated ROS is believed to be mitochondria. It has been estimated that nearly 2% of the total oxygen consumed by mitochondria leaks as partially reduced free oxygen radicals.[23] Another important source of intracellular ROS is the NADPH oxidase system which is known to be responsible for the respiratory burst microbicidal function in phagocytes. A low level of NADPH oxidase is also present in nonphagocytes and generates ROS with important roles with distinctive cellular functions including signal transduction.[24,25] Currently direct evidence is available linking ROS produced from NADPH oxidase to JNK activation in cells with various stimuli.[26,27]

Signaling Pathways Connecting ROS with JNK Activation

After the link between ROS/oxidative stress and JNK activation has been established, the main challenge is to elucidate the molecular mechanisms of ROS-mediated JNK activation. At present, a number of signaling pathways have been defined, one of them involves the MAP3K, ASK1. ASK1 is a ubiquitously expressed MAP3K that activates both JNK and p38 by phosphorylating and activating respective MAPKKs (JNKK1/MKK4, JNKK2/MKK7, MKK3 and MKK6).[28,29] ASK1 appears to be a specific target for ROS as oxidative stress readily activates ASK1 while suppresses MEKK1 by glutathionylation.[30] In the search of the molecular mechanism in ASK1 activation, Saitoh et al identified thioredoxin as the internal inhibitor of ASK1.[31] Thioredoxin is an important cellular redox regulatory protein[32] and such discovery thus directly links ROS and oxidative stress with ASK1 activation. The activity of ASK1 depends on the redox status of thioredoxin, and only the reduced form of thioredoxin, but not the oxidized form, is capable of binding to ASK1 and blocking its kinase activity. In the presence of ROS, the oxidized thioredoxin dissociates from ASK1 and therefore activates ASK1 by inducing oligomerization and subsequent phosphorylation of a critical threonine residue within the active loop of ASK1.[31,33,34] In addition, other function of thioredoxin on ASK1 has also been observed. A recent study demonstrated that thioredoxin is capable of promoting ASK1 ubiquitination and degradation to inhibit ASK1-mediated JNK activation and apoptosis.[35] Therefore, it is believed that the thioredoxin-ASK1 system serves as the molecular switch that converts redox signal to kinase activation. Additional mechanisms controlling ASK1 activation by ROS have also been proposed. For instance, a recent study revealed that ROS are able to activate ASK1 through its dissociation with the docking protein 14-3-3.[36] Protein phosphatase 5 (PP5) has been found to be another important internal inhibitor of ASK1. In response to ROS, it specifically binds to the activated form of ASK1 to block the ASK1-MKK4-JNK signaling cascade and to protect cell death.[37,38]

In addition to ASK1, other signaling pathways connecting ROS to JNK have also been suggested. There is evidence that Src kinase is an important redox-sensitive pathway for JNK activation.[39] It has been reported that either genetic disruption of Src or inhibition of Src kinase activity leads to specific suppression of JNK activation by H2O2, but not ERK and p38.[39] This finding is supported by another study in which Src was demonstrated as an upstream activator of JNK in endothelial cells treated with H2O2.[40] Currently, the exact signaling pathway from Src to JNK remains to be further elucidated. One clue is the role of Gab1, a docking protein downstream of epidermal growth factor receptor (EGFR) signaling. Upon H2O2 exposure, Gab1 is phosphorylated and binds to Src homology 2 (SH2)-containing protein-tyrosine phosphatase (SHP2), a critical step leading to specific activation of JNK, but not ERK or p38.[41] UV-induced JNK activation has also been found to follow a similar pathway.[42]

Another interesting signaling pathway in ROS-mediated JNK activation is the role of glutathione S-transferase Pi (GSTp). The monomeric form of GSTp binds to the c-terminal of JNK and suppress its kinase activity.[43] H2O2 treatment caused GSTp oligomerization and dissociation of the GSTp-JNK complex, leading to JNK activation.[44] It is worth mentioning that the effect of GSTp on JNK is independent of the MEKK1-MKK4 module, different from other signaling pathway as discussed above. Taken together, it is believed that ROS and oxidative are capable of modulating JNK activation via multiple signaling pathways.

JNK Activation in Oxidative Stress-Induced Cell Death

JNK Activation in Oxidative Stress-Induced Apoptosis

ROS and oxidative stress are known as apoptosis triggers and modulators.[45,46] Exogenous ROS such as H2O2 at moderate level induces apoptosis in many types of cells.[16] Endogenously produced ROS have also been found to be important in the apoptotic cell death process triggered by many other stimuli such as UV, ionizing radiation and cancer therapeutical drugs.[45] Since ROS themselves are not able to activate, but rather, to suppress redox-sensitive caspase activation,[16] ROS-induced apoptosis requires the participation of other cell death signaling pathways, including JNK. Extensive research on JNK has demonstrated that JNK serves as an important pro-apoptotic mechanism in oxidatively stressed cells. First, JNK is readily activated by ROS via distinct signaling pathways (as discussed earlier in this chapter). Second, suppression of JNK by either genetic or pharmacological approaches offers significant protection against apoptosis induced by ROS. For instance, ASK1-/- cells are resistant to H2O2-induced apoptosis.[34] Similar resistance was also found in cells with overexpression of dominant negative mutant of JNK.[47] Some antioxidants such as seleno-compounds are able to block H2O2-induced apoptosis via inhibition of ASK1-JNK.[48,49]

The next important question is how activated JNK leads to apoptosis. Currently many studies have suggested that mitochondria are the main site of action for JNK in apoptosis. The most conclusive evidence is from a study using primary murine embryonic fibroblasts with deletion of both JNK1 and JNK2.[50] Those cells are found to be resistant to UV-induced apoptosis due to defect in the mitochondrial death signaling pathway, including the failure to release cytochrome c. In support, it has been found that active JNK is able to induce the release of apoptogenic factors such as cytochrome c from isolated mitochondria in a cell-free assay.[51] Mitochondria translocation of JNK occurs in stressed cells,[52] and thus mitochondrially localized JNK provides the proximity to mitochondria-generated ROS and many apoptosis regulatory proteins such as Bcl-2 family proteins.[51,53,54] Bcl-2 family proteins are a major group of apoptosis regulatory factors closely associated with mitochondria. Currently there is accumulating evidence suggesting that the pro-apoptotic activity of JNK is executed via regulation of Bcl-2 family members. For instance, JNK can phosphorylate and inhibit the anti-apoptotic function of Bcl-2[55] and myeloid cell leukemia 1 (Mcl-1).[56] Moreover, JNK is capable of enhancing the function of pro-apoptotic Bcl-2 family members, via conformational changes and mitochondrial translocation of Bax,[57,58] phosphorylation of Bim and Bmf.[54,59] Therefore, it is believed that the Bcl-2 family proteins provide a molecular link between the JNK and the mitochondrial apoptotic machinery.

JNK Activation in Oxidative Stress-Induced Nonapoptotic Cell Death

It is believed that oxidative stress or ROS are capable of inducing either apoptosis or necrosis and the mode of cell death is determined by various factors, including the extent of oxidative stress, the cell type, and the nature of the cell signaling pathway elicited by ROS.[60,61] In comparison to oxidative stress-mediated apoptosis, the signaling pathway controlling oxidative stress-induced necrotic cell death is poorly understood. One question is whether JNK is also involved in ROS-induced necrotic or nonapoptotic cell death. If so, will JNK be activated via a similar mechanism as ROS-induced apoptotic cell death? A recent study in our laboratory has shed light on this mystery.[62] In mouse embryonic fibroblasts (MEF) with deletion of receptor-interacting protein (RIP) or tumor necrosis factor receptor-associated factor 2 (TRAF2), H2O2-induced JNK activation was significantly impaired and both *rip-/-* and *traf2-/-* MEF cells were resistant to H2O2-induced caspase-independent cell death. It is thus believed that both RIP and TRAF2, two critical effector molecules in TNF signaling, are important upstream regulators in H2O2-mediated JNK activation and nonapoptotic cell death. Moreover, such a JNK activation pathway is independent of ASK1-MKK4/MKK7, but rather relies on the direct protein-protein interaction between RIP/TRAF2 and JNK1, probably via colocalization in membrane lipid rafts upon H2O2 treatment. Therefore, data from our study provide a novel signaling pathway controlling ROS-induced JNK activation. Future work is required to identify the molecular targets of JNK implicated in ROS-induced necrotic/nonapoptotic cell death.

Distinct Roles of JNK1 and JNK2 in Oxidative Stress-Mediated Cell Death

Although the two ubiquitously expressed JNK1 and JNK2 function similarly in controlling of JNK activity, some recent studies have highlighted their functional differences in various aspects, including $CD8^+$ T cell activation,[63] UV-induced apoptosis,[64] and cell proliferation.[65] In H2O2-induced caspase-independent nonapoptotic cell death, distinct roles of JNK1 and JNK2 are also found, based on the observation that only *jnk1-/-* MEF cells are resistant to H2O2 cytotoxicity, while *jnk2-/-* cells are as sensitive as the wild-type MEF.[62] Consistently, only JNK1, but not JNK2, was found to directly interact with RIP upon H2O2 exposure (Shen et al, unpublished data). A very recent report by Liu et al found similar pattern of action: JNK1, but not JNK2, is essential for TNFα-induced apoptosis in MEF.[66] It remains to be determined whether the elevated level of ROS that often occurred in TNFα-treated cells contributes to the preferential activation of JNK1 in primary MEF or immortalized fibroblasts used in this study. The molecular mechanisms underlying such differences between the JNK1 and JNK2 in cell death are still elusive.

ROS and JNK Activation in TNF-Induced Apoptosis and Necrosis

TNF is a pleiotropic cytokine with critical regulatory roles in diverse cellular functions such as cell proliferation, differentiation and cell death. TNF elicits its biological effects via cell surface receptor TNFR1 and TNFR2. Three distinct signaling pathways in response to TNF have been intensively studied: (i) TNFR1-TRADD-FADD pathway to initiate caspase 8 activation to induce apoptotic cell death, (ii) TNFR1-TRADD-RIP/TRAF2 pathway to induce NF-κB activation to block apoptosis and promote cell survival, and (iii) TNFR1-TRADD-RIP/TRAF2 pathway to activate JNK.[67] In addition to the above three pathways, it has been well studied that ROS also function as important signaling molecules in TNF signaling.[22,68,69] In recent years, a number of important discoveries have further highlighted the critical role of ROS in TNF signaling, in particular in TNF-mediated NF-κB, JNK and cell death (both apoptotic and necrotic) pathways.

Differential Effects of ROS on TNF-Induced NF-κB and JNK Activation

Since the initial report by Shreck et al implied the involvement of ROS in TNFα-mediated NF-κB activation,[70] the exact effect of ROS on NF-κB has become a highly debatable issue and there are many reports either support or against this notion. The growing consensus is that ROS might not act as important NF-κB activators, at least in the TNF signaling pathways. Such an opinion was reinforced by a recent study with more conclusive evidence to dissociate the link between ROS and NF-κB activation.[71] In this study, it is demonstrated that N-acetyl-L-cysteine (NAC) and pyrrolidine dithiocarbamate (PDTC), the two antioxidants often used to block NF-κB activation, suppress TNF-mediated NF-κB pathway independent of their antioxidant function. Moreover, endogenous ROS produced through Rac/NADPH oxidase fail to activate NF-κB signaling, but instead reduce the magnitude of its activation.

On the other hand, there is emerging evidence that NF-κB has significant impact on ROS production and accumulation in TNF-treated cells. NF-κB acts as a suppressor of intracellular ROS formation in response to TNFα, based on the observations that elevated level of ROS was only found in TRAF2-TRAF5 double knockout or p65 knockout MEF cells, but not in wild-type cells upon TNF exposure.[72] Similar results were also found in IKKβ knockout MEF cells challenged with arsenic.[73] The remaining question to be answered is how NF-κB blocks TNF-stimulated ROS production and accumulation. One possible mechanism is that a number of key antioxidant enzymes are transcriptionally regulated by NF-κB and the deficiency of NF-κB will thus weaken the antioxidant defense system, leading to accumulation of ROS in the cell. Such a hypothesis is supported by data from a genome wide microarray analysis showing that upregulation of antioxidant enzymes via NF-κB pathway is crucial for elimination of ROS produced in TNFα-treated cells.[74] Moreover, a very recent study by Pham et al has identified a novel mechanism by which NF-κB executes its antioxidant function.[75] NF-κB controls the expression of ferritin heavy chain (FHC), the primary iron storage factor highly capable of suppressing ROS accumulation through iron sequestration, to prevent sustained JNK activation and, thereby, apoptosis triggered by TNFα.

In contrast to often contradictory effect of ROS on NF-κB activation, the importance of ROS in TNF-mediated JNK activation has been consistently demonstrated in various cell systems.[11,76] It is now recognized that there are two phases of JNK activation mediated by two different activation pathways: the earlier and transient activation of JNK is mediated by TRAF2,[77] while the delayed and persistent activation of JNK is mediated by ROS.[72,78] Intriguingly, being opposite to the antioxidant function of NF-κB, JNK activation contributes to TNF-stimulated ROS production. Upon TNFα stimulation, elevated ROS level was only found in wild-type mouse fibroblasts in which the NF-κB pathway is abolished by overexpression of ΔN-IκBα, but not in *jnk-/-*(ΔN-Iκα) cells.[78] Although it is not known how JNK promotes ROS production, it appears that there is a positive feedback loop between JNK activation and ROS production, and both work together contributing to TNFα-induced cell death.

ROS as the Key Mediator for the Interplay between NF-κB and JNK Activation

Two earlier studies reported that the NF-κB and the JNK signaling pathway are functionally interlinked.[79,80] In normal cells TNFα induces an earlier and transient JNK activation, while in cells that are deficient in NF-κB activation, JNK activation is enhanced and sustained. More importantly, these studies also demonstrated that the anti-apoptotic function of NF-κB is, at least in part, achieved through its ability to suppress the prolonged JNK activation. Two NF-κB target genes, XIAP[79] and GADD45β,[80,81] have been identified as the mediators regulating the cross-talks between these two key signaling pathways.

Recently ROS have emerged as another potential mediator for the interplay between NF-κB and JNK in cells treated with TNF.[82,83] Most recently, Kamata et al demonstrated that ROS promote TNF-induced cell death and sustained JNK activation by inhibiting MAP kinase phosphatases.[91] Therefore, TNF-induced ROS production mediates the sustained JNK activation and cell death, and, conversely, that TNF-induced NF-κB activation prevents ROS production and accumulation. Moreover, the sequential events of NF-κB inhibition, ROS accumulation (oxidative stress) and sustained JNK activation seem to be the key steps in TNF-induced cell death. ROS and oxidative stress may occupy a unique position in apoptotic cell death induced by TNFα or other stimuli capable of activating NF-κB and JNK. As most studies discussed above are resulted from using mouse fibroblasts, it remains to be further investigate whether the mediating role of ROS between JNK and NF-κB can be extended to other cells such as human cancer cells. An earlier report revealed that ROS potentiate JNK activation and apoptosis in oncogenically transformed NIH 3T3 cells, suggesting such a possibility.[84] In this regard, we should also be cautious of using antioxidants in combination with cancer chemotherapy as the removal of ROS in fact will lessen the desired cell death in cancer.

Role of ROS-Mediated JNK Activation in TNF-Induced Necrotic Cell Death

Depending on the cell type and the presence or absence of other signaling pathways, JNK can be pro-apoptotic, anti-apoptotic or no role in TNF-mediated apoptosis.[3,85,86] One common understanding is that sustained JNK activation mediated by ROS enhances TNFα-induced apoptosis, most probably through the mitochondria-dependent apoptosis pathway. One recent report by Deng et al strengthened the pro-apoptotic function of JNK by showing that sustained JNK activation induces caspase 8-independent cleavage of Bid, which then induces preferential release of Smac from mitochondria to promote apoptosis.[87]

One of the important findings in the study of cell death is that cell death ligands (FasL and TNFα) are able to induce caspase-independent or necrotic cell death. Such a phenomenon was first observed in one particular cell line, murine L929 fibrosarcoma cells.[88,89] Interestingly, endogenously generated ROS have been implicated in TNFα-induced necrosis in L929 cells.[69] Subsequent studies have provided convincing evidence that TNFα-induced necrosis is also present in other cells such as fibroblasts.[72,78,90] In the search of the molecular mechanism controlling TNFα-induced necrotic cell death, most studies suggested a role of ROS and oxidative stress. For instance, Lin et al reported that TNFα-induced nonapoptotic cell death was only evident in wild-type MEF cells with higher level of ROS, but not in *rip-/-*, *traf2-/-* MEF without ROS accumulation after TNFα treatment.[90] Being consistent with our findings that exogenous applied ROS induce necrotic cell death via JNK activation,[62] JNK has also been proposed as a critical mediator in TNFα-induced necrotic cell death. In NF-κB knockout MEF (*p65-/-*), treatment with TNFα promotes ROS production that mediates sustained JNK activation and necrotic cell death.[72,91] More recently, it is further demonstrated that JNK can potentiate TNFα-stimulated necrosis by increasing the production of ROS.[78,91] Those findings thus suggest the dual functions of JNK: JNK acts as a critical mediator in ROS-induced both apoptosis and nonapoptotic cell death.

Summary and Perspectives

The importance of ROS and JNK activation in cell death has been realized in the recent years. It is believed that ROS-mediated JNK activation is a critical component deciding the fate of cells in response to various stress stimuli, including TNFα. Such an understanding regarding ROS and JNK activation may have broad practical implications in biology and treatment of diseases. One possibility is to specifically enhance ROS-dependent potentiation of JNK and thus sensitize cancer cells to cancer chemotherapeutics. On the other hand, concerns have been raised regarding potential side effects due to oxidative stress following NF-κB inhibition, a common strategy in treatment of inflammatory diseases.[82,83] Further more, cautions should be given for using antioxidants in the course of cancer chemotherapy as the removal of ROS will in fact compromise the effectiveness of the chemotherapeutics. At present, many intriguing questions about the function of ROS and JNK activation remain to be further explored. For instance, what is the underlying mechanism for the promoting effect of ROS and prolonged JNK activation in cell death? Do ROS and sustained JNK activation work through similar machinery in apoptotic and necrotic cell death? What is the down stream target(s) of the prolonged JNK activation in cell death? All these should be the focus of future studies in this rather exciting field.

Acknowledgement

The authors would like to thank Dr. Yong Lin for his valuable comments. H.M. Shen is supported by an academic research fund from the National University of Singapore.

References

1. Davis RJ. Signal transduction by the JNK group of MAP kinases. Cell 2000; 103:239-252.
2. Minden A, Karin M. Regulation and function of the JNK subgroup of MAP kinases. Biochim Biophys Acta 1997; 1333:F85-104.
3. Lin A. Activation of the JNK signaling pathway: Breaking the brake on apoptosis. Bioessays 2003; 25:17-24.
4. Weston CR, Davis RJ. The JNK signal transduction pathway. Curr Opin Genet Dev 2002; 12:14-21.
5. Xia Y, Makris C, Su B et al. MEK kinase 1 is critically required for c-Jun N-terminal kinase activation by proinflammatory stimuli and growth factor-induced cell migration. Proc Natl Acad Sci USA 2000; 97:5243-5248.
6. Yujiri T, Sather S, Fanger GR et al. Role of MEKK1 in cell survival and activation of JNK and ERK pathways defined by targeted gene disruption. Science 1998; 282:1911-1914.
7. Matsuzawa A, Nishitoh H, Tobiume K et al. Physiological roles of ASK1-mediated signal transduction in oxidative stress- and endoplasmic reticulum stress-induced apoptosis: Advanced findings from ASK1 knockout mice. Antioxid Redox Signal 2002; 4:415-425.
8. Sies H. Oxidative stress: Oxidants and antioxidants. Exp Physiol 1997; 82:291-295.
9. Droge W. Free radicals in the physiological control of cell function. Physiol Rev 2002; 82:47-95.
10. Martindale JL, Holbrook NJ. Cellular response to oxidative stress: Signaling for suicide and survival. J Cell Physiol 2002; 192:1-15.
11. Torres M. Mitogen-activated protein kinase pathways in redox signaling. Front Biosci 2003; 8:d369-91-d369-d391.
12. Mittler R, Berkowitz G. Hydrogen peroxide, a messenger with too many roles? Redox Rep 2001; 6:69-72.
13. Rhee SG, Bae YS, Lee SR et al. Hydrogen peroxide: A key messenger that modulates protein phosphorylation through cysteine oxidation. Sci STKE 2000.
14. Sata N, Klonowski-Stumpe H, Han B et al. Menadione induces both necrosis and apoptosis in rat pancreatic acinar AR4-2J cells. Free Radic Biol Med 1997; 23:844-850.
15. Ma X, Du J, Nakashima I et al. Menadione biphasically controls JNK-linked cell death in leukemia Jurkat T cells. Antioxid Redox Signal 2002; 4:371-378.
16. Ueda S, Masutani H, Nakamura H et al. Redox control of cell death. Antioxid Redox Signal 2002; 4:405-414.
17. Ding M, Li J, Leonard SS et al. Differential role of hydrogen peroxide in UV-induced signal transduction. Mol Cell Biochem 2002; 234-235:81-90.
18. Chan WH, Wu CC, Yu JS. Curcumin inhibits UV irradiation-induced oxidative stress and apoptotic biochemical changes in human epidermoid carcinoma A431 cells. J Cell Biochem 2003; 90:327-338.

19. Lee SA, Dritschilo A, Jung M. Role of ATM in oxidative stress-mediated c-Jun phosphorylation in response to ionizing radiation and CdCl2. J Biol Chem 2001; 276:11783-11790.
20. Mansat-de MV, Bezombes C, Quillet-Mary A et al. Implication of radical oxygen species in ceramide generation, c-Jun N-terminal kinase activation and apoptosis induced by daunorubicin. Mol Pharmacol 1999; 56:867-874.
21. Yu C, Rahmani M, Dent P et al. The hierarchical relationship between MAPK signaling and ROS generation in human leukemia cells undergoing apoptosis in response to the proteasome inhibitor Bortezomib. Exp Cell Res 2004; 295:555-566.
22. Garg AK, Aggarwal BB. Reactive oxygen intermediates in TNF signaling. Mol Immunol 2002; 39:509-517.
23. Ames BN, Shigenaga MK, Hagen TM. Oxidants, antioxidants, and the degenerative diseases of aging. Proc Natl Acad Sci USA 1993; 90:7915-7922.
24. Lambeth JD. NOX enzymes and the biology of reactive oxygen. Nat Rev Immunol 2004; 4:181-189.
25. Werner E. GTPases and reactive oxygen species: Switches for killing and signaling. J Cell Sci 2004; 117:143-153.
26. Xu YC, Wu RF, Gu Y et al. Involvement of TRAF4 in oxidative activation of c-Jun N-terminal kinase. J Biol Chem 2002; 277:28051-28057.
27. Gu Y, Xu YC, Wu RF et al. TNFalpha activates c-Jun amino terminal kinase through p47(phox). Exp Cell Res 2002; 272:62-74.
28. Wang XS, Diener K, Jannuzzi D et al. Molecular cloning and characterization of a novel protein kinase with a catalytic domain homologous to mitogen-activated protein kinase kinase kinase. J Biol Chem 1996; 271:31607-31611.
29. Ichijo H, Nishida E, Irie K et al. Induction of apoptosis by ASK1, a mammalian MAPKKK that activates SAPK/JNK and p38 signaling pathways. Science 1997; 275:90-94.
30. Cross JV, Templeton DJ. Oxidative stress inhibits MEKK1 by site-specific glutathionylation in the ATP-binding domain. Biochem J 2004; 381:675-683.
31. Saitoh M, Nishitoh H, Fujii M et al. Mammalian thioredoxin is a direct inhibitor of apoptosis signal-regulating kinase (ASK) 1. EMBO J 1998; 17:2596-2606.
32. Nordberg J, Arner ES. Reactive oxygen species, antioxidants, and the mammalian thioredoxin system. Free Radic Biol Med 2001; 31:1287-1312.
33. Gotoh Y, Cooper JA. Reactive oxygen species- and dimerization-induced activation of apoptosis signal-regulating kinase 1 in tumor necrosis factor-alpha signal transduction. J Biol Chem 1998; 273:17477-17482.
34. Tobiume K, Matsuzawa A, Takahashi T et al. ASK1 is required for sustained activations of JNK/p38 MAP kinases and apoptosis. EMBO Rep 2001; 2:222-228.
35. Liu Y, Min W. Thioredoxin promotes ASK1 ubiquitination and degradation to inhibit ASK1-mediated apoptosis in a redox activity-independent manner. Circ Res 2002; 90:1259-1266.
36. Goldman EH, Chen L, Fu H. Activation of apoptosis signal-regulating kinase 1 by reactive oxygen species through dephosphorylation at serine 967 and 14-3-3 dissociation. J Biol Chem 2004; 279:10442-10449.
37. Morita K, Saitoh M, Tobiume K et al. Negative feedback regulation of ASK1 by protein phosphatase 5 (PP5) in response to oxidative stress. EMBO J 2001; 20:6028-6036.
38. Zhou G, Golden T, Aragon IV et al. Ser/Thr protein phosphatase 5 inactivates hypoxia-induced activation of an apoptosis signal-regulating kinase 1/MKK-4/JNK signaling cascade. J Biol Chem 2004; 279:46595-46605.
39. Yoshizumi M, Abe J, Haendeler J et al. Src and Cas mediate JNK activation but not ERK1/2 and p38 kinases by reactive oxygen species. J Biol Chem 2000; 275:11706-11712.
40. Chen K, Vita JA, Berk BC et al. c-Jun N-terminal kinase activation by hydrogen peroxide in endothelial cells involves SRC-dependent epidermal growth factor receptor transactivation. J Biol Chem 2001; 276:16045-16050.
41. Holgado-Madruga M, Wong AJ. Gab1 is an integrator of cell death versus cell survival signals in oxidative stress. Mol Cell Biol 2003; 23:4471-4484.
42. Sun Y, Yuan J, Liu H et al. Role of Gab1 in UV-induced c-Jun NH2-terminal kinase activation and cell apoptosis. Mol Cell Biol 2004; 24:1531-1539.
43. Wang T, Arifoglu P, Ronai Z et al. Glutathione S-transferase P1-1 (GSTP1-1) inhibits c-Jun N-terminal kinase (JNK1) signaling through interaction with the C terminus. J Biol Chem 2001; 276:20999-21003.
44. Adler V, Yin Z, Fuchs SY et al. Regulation of JNK signaling by GSTp. EMBO J 1999; 18:1321-1334.
45. Chandra J, Samali A, Orrenius S. Triggering and modulation of apoptosis by oxidative stress. Free Radic Biol Med 2000; 29:323-333.

46. Carmody RJ, Cotter TG. Signalling apoptosis: A radical approach. Redox Rep 2001; 6:77-90.
47. Li Y, Arita Y, Koo HC et al. Inhibition of c-Jun N-terminal kinase pathway improves cell viability in response to oxidant injury. Am J Respir Cell Mol Biol 2003; 29:779-783.
48. Yoon SO, Kim MM, Park SJ et al. Selenite suppresses hydrogen peroxide-induced cell apoptosis through inhibition of ASK1/JNK and activation of PI3-K/Akt pathways. FASEB J 2002; 16:111-113.
49. Yoshizumi M, Kogame T, Suzaki Y et al. Ebselen attenuates oxidative stress-induced apoptosis via the inhibition of the c-Jun N-terminal kinase and activator protein-1 signalling pathway in PC12 cells. Br J Pharmacol 2002; 136:1023-1032.
50. Tournier C, Hess P, Yang DD et al. Requirement of JNK for stress-induced activation of the cytochrome c-mediated death pathway. Science 2000; 288:870-874.
51. Aoki H, Kang PM, Hampe J et al. Direct activation of mitochondrial apoptosis machinery by c-Jun N-terminal kinase in adult cardiac myocytes. J Biol Chem 2002; 277:10244-10250.
52. Kharbanda S, Saxena S, Yoshida K et al. Translocation of SAPK/JNK to mitochondria and interaction with Bcl-x(L) in response to DNA damage. J Biol Chem 2000; 275:322-327.
53. Brichese L, Cazettes G, Valette A. JNK is Associated with Bcl-2 and PP1 in mitochondria: Paclitaxel induces its activation and its association with the phosphorylated form of Bcl-2. Cell Cycle 2004; 3.
54. Putcha GV, Le S, Frank S et al. JNK-mediated BIM phosphorylation potentiates BAX-dependent apoptosis. Neuron 2003; 38(19):899-914.
55. Yamamoto K, Ichijo H, Korsmeyer SJ. BCL-2 is phosphorylated and inactivated by an ASK1/Jun N-terminal protein kinase pathway normally activated at G(2)/M. Mol Cell Biol 1999; 19:8469-8478.
56. Inoshita S, Takeda K, Hatai T et al. Phosphorylation and inactivation of myeloid cell leukemia 1 by JNK in response to oxidative stress. J Biol Chem 2002; 277:43730-43734.
57. Tsuruta F, Sunayama J, Mori Y et al. JNK promotes Bax translocation to mitochondria through phosphorylation of 14-3-3 proteins. EMBO J 2004; 23:1889-1899.
58. Lei K, Nimnual A, Zong WX et al. The Bax subfamily of Bcl2-related proteins is essential for apoptotic signal transduction by c-Jun NH(2)-terminal kinase. Mol Cell Biol 2002; 22:4929-4942.
59. Lei K, Davis RJ. JNK phosphorylation of Bim-related members of the Bcl2 family induces Bax-dependent apoptosis. Proc Natl Acad Sci USA 2003; 100:2432-2437.
60. Troyano A, Sancho P, Fernandez C et al. The selection between apoptosis and necrosis is differentially regulated in hydrogen peroxide-treated and glutathione-depleted human promonocytic cells. Cell Death Differ 2003; 10:889-898.
61. Gardner AM, Xu FH, Fady C et al. Apoptotic vs. nonapoptotic cytotoxicity induced by hydrogen peroxide. Free Radic Biol Med 1997; 22:73-83.
62. Shen HM, Lin Y, Choksi S et al. Essential roles of receptor-interacting protein and TRAF2 in oxidative stress-induced cell death. Mol Cell Biol 2004; 24:5914-5922.
63. Conze D, Krahl T, Kennedy N et al. c-Jun NH(2)-terminal kinase (JNK)1 and JNK2 have distinct roles in CD8(+) T cell activation. J Exp Med 2002; 195:811-823.
64. Hochedlinger K, Wagner EF, Sabapathy K. Differential effects of JNK1 and JNK2 on signal specific induction of apoptosis. Oncogene 2002; 21:2441-2445.
65. Sabapathy K, Hochedlinger K, Nam SY et al. Distinct roles for JNK1 and JNK2 in regulating JNK activity and c-Jun-dependent cell proliferation. Mol Cell 2004; 15:713-725.
66. Liu J, Minemoto Y, Lin A. c-Jun N-terminal protein kinase 1 (JNK1), but not JNK2, is essential for tumor necrosis factor alpha-induced c-Jun kinase activation and apoptosis. Mol Cell Biol 2004; 24:10844-10856.
67. Chen G, Goeddel DV. TNF-R1 signaling: A beautiful pathway. Science 2002; 296:1634-1635.
68. Goossens V, De Vos K, Vercammen D et al. Redox regulation of TNF signaling. Biofactors 1999; 10:145-156.
69. Fiers W, Beyaert R, Declercq W et al. More than one way to die: Apoptosis, necrosis and reactive oxygen damage. Oncogene 1999; 18:7719-7730.
70. Schreck R, Rieber P, Baeuerle PA. Reactive oxygen intermediates as apparently widely used messengers in the activation of the NF-kappa B transcription factor and HIV-1. EMBO J 1991; 10:2247-2258.
71. Hayakawa M, Miyashita H, Sakamoto I et al. Evidence that reactive oxygen species do not mediate NF-kappaB activation. EMBO J 2003; 22:3356-3366.
72. Sakon S, Xue X, Takekawa M et al. NF-kappaB inhibits TNF-induced accumulation of ROS that mediate prolonged MAPK activation and necrotic cell death. EMBO J 2003; 22:3898-3909.
73. Chen F, Castranova V, Li Z et al. Inhibitor of nuclear factor kappaB kinase deficiency enhances oxidative stress and prolongs c-Jun NH2-terminal kinase activation induced by arsenic. Cancer Res 2003; 63:7689-7693.

74. Sasazuki T, Okazaki T, Tada K et al. Genome wide analysis of TNF-inducible genes reveals that antioxidant enzymes are induced by TNF and responsible for elimination of ROS. Mol Immunol 2004; 41:547-551.
75. Pham CG, Bubici C, Zazzeroni F et al. Ferritin heavy chain upregulation by NF-kappaB inhibits TNFalpha-induced apoptosis by suppressing reactive oxygen species. Cell 2004; 119:529-542.
76. Takeda K, Matsuzawa A, Nishitoh H et al. Roles of MAPKKK ASK1 in stress-induced cell death. Cell Struct Funct 2003; 28:23-29.
77. Lamb JA, Ventura JJ, Hess P et al. JunD mediates survival signaling by the JNK signal transduction pathway. Mol Cell 2003; 11:1479-1489.
78. Ventura JJ, Cogswell P, Flavell RA et al. JNK potentiates TNF-stimulated necrosis by increasing the production of cytotoxic reactive oxygen species. Genes Dev 2004; 18:2905-2915.
79. Tang G, Minemoto Y, Dibling B et al. Inhibition of JNK activation through NF-kappaB target genes. Nature 2001; 414:313-317.
80. De SE, Zazzeroni F, Papa S et al. Induction of gadd45beta by NF-kappaB downregulates pro-apoptotic JNK signalling. Nature 2001; 414:308-313.
81. Papa S, Zazzeroni F, Bubici C et al. Gadd45 beta mediates the NF-kappa B suppression of JNK signalling by targeting MKK7/JNKK2. Nat Cell Biol 2004; 6:146-153.
82. Zhang Y, Chen F. Reactive oxygen species (ROS), troublemakers between nuclear factor-kappaB (NF-kappaB) and c-Jun NH(2)-terminal kinase (JNK). Cancer Res 2004; 64:1902-1905.
83. Nakano H. Signaling crosstalk between NF-kappaB and JNK. Trends Immunol 2004; 25:402-405.
84. Benhar M, Dalyot I, Engelberg D et al. Enhanced ROS production in oncogenically transformed cells potentiates c-Jun N-terminal kinase and p38 mitogen-activated protein kinase activation and sensitization to genotoxic stress. Mol Cell Biol 2001; 21:6913-6926.
85. Varfolomeev EE, Ashkenazi A. Tumor necrosis factor: An apoptosis JuNKie? Cell 2004; 116(20):491-497.
86. Liu ZG, Hsu H, Goeddel DV et al. Dissection of TNF receptor 1 effector functions: JNK activation is not linked to apoptosis while NF-kappaB activation prevents cell death. Cell 1996; 87:565-576.
87. Deng Y, Ren X, Yang L et al. A JNK-dependent pathway is required for TNFalpha-induced apoptosis. Cell 2003; 115:61-70.
88. Vercammen D, Beyaert R, Denecker G et al. Inhibition of caspases increases the sensitivity of L929 cells to necrosis mediated by tumor necrosis factor. J Exp Med 1998; 187:1477-1485.
89. Vercammen D, Brouckaert G, Denecker G et al. Dual signaling of the Fas receptor: Initiation of both apoptotic and necrotic cell death pathways. J Exp Med 1998; 188:919-930.
90. Lin Y, Choksi S, Shen HM et al. Tumor necrosis factor-induced nonapoptotic cell death requires receptor-interacting protein-mediated cellular reactive oxygen species accumulation. J Biol Chem 2004; 279:10822-10828.
91. Kamata H, Honda S, Maeda S et al. Reactive oxygen species promote TNFalpha-induced death and sustained JNK activation by inhibiting MAP kinase phosphatases. Cell 2005; 120:649-61.

CHAPTER 8

JNK as a Therapeutic Target

Brydon L. Bennett* and Yoshitaka Satoh

Abstract

Promising drug targets have activities that are highly correlated with disease and where interference of activity results in desired changes in functional markers of disease. Increased JNK activity has been measured in biopsy samples from a range of human diseases. Inhibition of JNK activity using knockout technology, dominant negative decoys, and small molecule or peptide inhibitors has confirmed the importance of JNK in promoting a pathological phenotype in many animal models of disease. The first JNK inhibitors are now entering clinical trials to determine safety and efficacy limits in humans.

Therapeutic Potential

JNK is a stress-activated protein kinase.[1-3] When we consider that cellular and molecular stress includes a myriad of physical, chemical and biological insults it is understandable how frequently activated JNK has been found to be associated with pathology. Examples of molecular stressors include radiation, acidosis, hypoxia, burn, osmotic shock, cytokine and chemokine stimulation, growth factors, antigen/allergen challenge, and infection. Some important molecular mechanisms whereby JNK transduces the stress signal through the cell have been elucidated. These include the classical pathway of c-Jun phosphorylation, AP-1 transcription factor activation and the subsequent transcription of a large set of response genes including c-Jun itself. Recall that c-Jun was initially characterized as a proto-oncogene suggesting a role in cellular transformation and proliferation. The expression of AP-1 regulated genes may be magnified by the action of JNK on AU-rich elements of the untranslated mRNA to stabilize mRNA and increase protein translation.[4] Other JNK substrates have been identified including regulatory proteins such as IRS-1 that control insulin signaling and by which JNK phosphorylation may fundamentally promote insulin resistance.[5] The pro- and anti-apoptotic Bcl family of proteins have also been proposed as JNK substrates.[6,7] If so, this would provide a molecular link to the observed requirement for JNK in promoting mitochondrial cytochrome c release and apoptosis.

By integrating our mechanistic observations with the biology of disease it is apparent that JNK inhibitors have broad clinical potential as therapeutics ranging from organ failure due to ischemia-reperfusion injury (stroke, myocardial infarction, traumatic injury, surgery), neurodegenerative diseases (Alzheimer's, Parkinson's, cognitive dysfunction), autoimmune and inflammatory diseases (arthritis, asthma, psoriasis, graft rejection), metabolic disease (obesity and type 2 diabetes), and proliferative disorders (fibrosis, cancer). It will take time to chart all these areas but the clinical burden exemplified by these diseases is obvious.

JNK inhibitors represent an innovative therapeutic approach for the treatment of a range of clinical indications. But it is only fair that we also speculate on potential side-effects of JNK

*Corresponding Author: Brydon L. Bennett—Inflammation Research, Celgene, 4550 Towne Centre Court, San Diego, California 92121, U.S.A. Email: bbennett@celgene.com

targeted therapy. Perhaps the most obvious is the potential for myelosuppression and immunosuppression. The former because JNK is believed to play a role in hematopoiesis[8] and the latter because of the necessity of JNK for maturation and differentiation of CD4$^+$ and CD8$^+$ T-cells.[9] Note that this potential toxicity may also be a therapeutic benefit in certain situations. As detailed in another chapter of this book, it has been shown that the JNK homolog in Drosophila, ***basket***, is essential for dorsal closure of the larval embryo. This is noteworthy because this process in invertebrates is believed to be analogous to wound healing in vertebrates.[10,11] Therefore it can be postulated that high levels of systemic JNK inhibitor may delay or disrupt organ repair particularly when involving epithelial cells. A third potential mechanistic side-effect of JNK inhibitors is susceptibility to viral infection. This hypothesis is built on the observation that many viruses express Bcl related proteins as a direct means of preventing apoptosis of the host cell, thereby allowing viral replication.[12] If JNK is involved in promoting apoptosis of virally infected cells, it is possible that inhibition of JNK will work in favor of the infectious virus.

Much of our enthusiasm for a role of JNK in disease has been accrued from studies using knockout animals. Before we move into a brief description of JNK drug discovery it is worth commenting that the pharmacodynamics of a JNK inhibitor are unlikely to result in a phenotype identical to that reported for JNK knock-out animals. First, most JNK inhibitors inhibit all 3 JNK isoforms, but we do not yet have a genetic equivalent because *Jnk*1$^{-/-}$ *Jnk2*$^{-/-}$ double knock-out animals are embryonic lethal. Second, most knockouts ablate expression of the entire protein rather than negating only its catalytic activity as expected of small molecule inhibitors. Therefore, noncatalytic functions of JNK may be preserved despite the binding of a small molecule inhibitor. Finally, the kinetics of small molecule binding and the pharmacodynamics that determine tissue active concentrations should make JNK inhibition both dynamic and manipulatable. For example, 80% inhibition of activated JNK for 16 hours per day may have therapeutic benefit without compromising basal or housekeeping JNK functions. For these reasons, only clinical trials with candidate molecules will accurately resolve these questions.

JNK Drug Discovery

Biologically interactive proteins that contribute to pathological outcomes are attractive drug targets. Enzymes represent an exploitable subset of targets because they possess allosteric and or catalytic active sites that can be effectively engaged with a small molecule (MW < 700 Daltons). Protein kinases are validated drug targets as proven by the commercialization of drugs such as Eril (*fasudil*), Gleevec (*imatinib*), Iressa (*gefitinib*) and Tarceva (*erlotinib*).[13] A distinct advantage in optimizing inhibitors of protein kinases is that these enzymes can usually be expressed as active recombinant proteins, which is essential for establishing biochemical screens and generating crystal structures. These two techniques are the platform for what is termed 'rational' drug discovery – the current industry paradigm for identifying and optimizing small molecule inhibitors to be used as therapeutics. The design and validation of biochemical assays for JNK has been described in detail.[14] In essence, active recombinant JNK is mixed with c-Jun and ATP in the presence or absence of known concentrations of test compound. JNK activity, or inhibition thereof, is monitored by the amount of phosphorylated c-Jun formed. Although early screens used incorporation of radiolabeled phosphate (^{32}P or ^{33}P), most detection systems now use fluorescent conjugated antibodies specific for phosphorylated epitopes on c-Jun. In theory, this assay design can identify inhibitors of any docking site that might suppress catalytic activity. In reality, as illustrated in section 3, all small molecule inhibitors of JNK disclosed too date appear to bind in the active site. This is in part a result of the simplified biochemical assay described above, devoid of associated proteins (e.g., scaffolds) that might modulate activity in vivo and be sensitive to inhibitors in vitro. It is also due to a fundamental desire to understand mechanism, that biases our efforts toward the safe harbor of ATP competitive kinetics and the architecture of the kinase active site as revealed by crystallography. Allosteric sites on JNK that can be targeted with small molecules have not been described, however it is noteworthy that non ATP competitive inhibitors are in development for the related kinases p38 MAPK, IKKβ and MEK.[15,16] Such alternative binding modes may have distinct advantages over classical ATP-competitive inhibitors. Firstly, as reversible binders, ATP competitive molecules are directly affected by the high concentration of ATP in the cell (estimated to range from 0.2-5 mM) requiring low nanomolar Ki values to give pharmacologically acceptable (typically < 1 μM) IC_{50} values in vivo. Secondly, since all protein kinases fundamentally perform the same catalytic reaction, the active sites are structurally conserved across the kinase family with highly related kinases often differing in only a handful of amino acids. Furthermore, architectural similarity in the active site may be built around different amino acids leading to unexpectedly potent compound interactions with kinases that appear distantly related according to sequence homology.[17] This increases the challenge for identifying inhibitors with the desired selectivity profile and has highlighted the utility of crystal structures to identify individual molecular features that can be targeted specifically by the chemist. The crystal structure of JNK3 was first described by Vertex[18] and the details of SP600125 and JIP binding to JNK3 and JNK1 have also been published[19,20] (please also see Fig. 1). Crystal structures for JNK2 have not been disclosed but are believed to exist. Investigation of the active sites of the JNK isoforms reveals almost complete identity between JNK2 and JNK3 but there are a few notable amino acid differences in JNK1, including V53I, V85I, D87E, S125G and L144I (numbered according to JNK3). To date the majority of JNK inhibitors have been pan-inhibitors of all 3 isoforms, however at the American Chemical Society meeting in Anaheim, CA in April 2004, Amgen presented a benzoxazepinone compound with IC_{50} values of 11, 0.11, and 0.047 μM for JNK1, 2 and 3 respectively. Taken together the data suggest that some degree of isoform selectivity may be obtainable between JNK1 and JNK2/3 when targeting the active site. This could be therapeutically advantageous with respect to specificity and safety since knockout studies have revealed isoform specific roles in disease models as described in a previos chapter.

A number of cell assays useful for measuring JNK activity have been described.[14,21] The JNK directed phosphorylation sites on c-Jun, namely serines' -63 and -73 appear to be highly specific for JNK which means that quantitation of ser-63/73 phosphorylation is an accurate biomarker for assessing JNK inhibitor potency in cells. Unfortunately, we have found that the commercially available phosphospecific antibodies have only modest binding affinity and frequently exhibit nonspecific binding. Therefore while these reagents are useful in the context of western blots where c-Jun can be identified by molecular weight they are more troublesome in intact cell/tissue formats especially where high-throughput analysis is desired. A number of additional substrates for JNK have been identified including ATF2, Bcl family proteins, IRS-1 and ShcA but specificity for JNK has not been demonstrated. Functional markers of JNK activity include a host of c-Jun/AP-1 regulated genes[22] but here again it is apparent that these genes are regulated by multiple transcription factors and signaling pathways and an exclusively JNK regulated gene may not exist. Instead, analysis of the selectivity profile of the inhibitor coupled to an understanding of the regulation of specific genes relevant to the therapeutic area may be necessary to select the appropriate functional marker.

Using the technology and assays just described, a rapidly growing number of JNK inhibitors are being identified. As we shall see in the following section, small molecule inhibitors have provided us with pharmacologic evidence on the importance of JNK in animal models of disease and are starting to be evaluated in the clinic.

Small Molecule Inhibitors

Based on published disclosures, the development of JNK inhibitors appears to be focused on the optimization of small molecules that act within the kinase active site, although as we shall see later, other strategies may also be viable. A description of some early JNK inhibitors has been presented.[23] The first widely used JNK inhibitor was SP600125 (anthra[1,9-cd]pyrazol-6(2H)-one) from Signal Pharmaceuticals/Celgene, an ATP competitive inhibitor of JNK1, 2 and 3 with a Ki of 100-200 nM.[24] The advantages of this compound

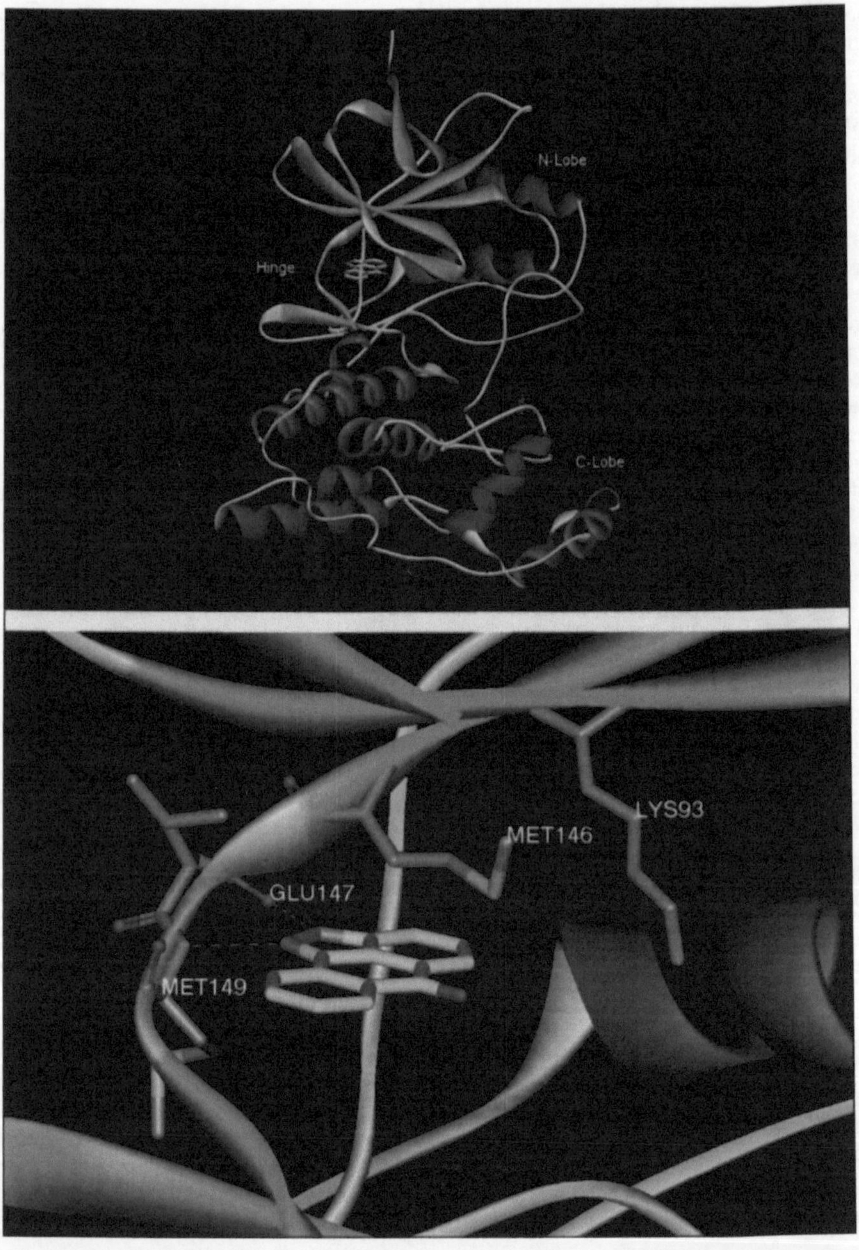

Figure 1. Structure of JNK3 with SP600125 bound in the active site (19). A) Ribbon diagram showing structure of JNK3 kinase with SP600125 (yellow). β-sheet motifs are in blue and α-helices are in red. The N-lobe, C-lobe, and hinge region of the kinase are indicated. SP600125, shown in yellow, binds at the ATP-binding site of JNK3. B) Close-up view showing SP600125 binding interactions in ATP-binding site. Labeled residues Met149 and Glu147 are in the hinge region and are points of hydrogen bond interaction with nitrogen atoms (purple) of SP600125. Met146 is the 'gatekeeper' residue, and Lys93 is the conserved catalytic lysine. SP600125 is a small planar molecule. Optimized inhibitors may be expected to fill additional space and form additional interactions with active site residues.

were its specificity of action and its selectivity against the related MAP kinases ERK and p38. Consequently SP600125 has been used as a pharmacologic tool in over 300 journal articles. Consistent with knock-out and biochemical studies, SP600125 has been shown to inhibit endotoxin induced TNFα expression,[24] suppress bone degeneration in models of arthritis,[25,26] inhibit acute T-cell inflammation and liver injury,[27] decrease allergen induced airway inflammation and remodeling[28] enhance survival of dopaminergic neurons in the MPTP model,[29] and protect from ischemia-reperfusion injury.[30] However, SP600125 does inhibit multiple other protein kinases,[31] may have additional cell activity,[32] and lacks the physical chemical properties required of a clinical compound.

At the American Chemical Society meeting in 2003, Serono described a novel benzothiazol-2-yl acetonitrile derivative JNK inhibitor, AS601245, whose characterization was later published in full.[33] This molecule was a pan inhibitor of JNK isoforms, with potency equivalent to SP600125 but with an improved kinase selectivity profile. Consistent with a role for JNK in stress-induced apoptosis, AS601245 exhibited significant protection in a gerbil model of cerebral ischemia and a rat model of myocardial infarction with concomitant inhibition of c-Jun phosphorylation in vivo.[33,34] Although once in late stage preclinical development, AS601245 is now used as a proof of concept reagent and benchmark for second generation compounds. At the 2nd Anglo-Swedish Medicinal Chemistry Symposium (March 2005) Serono announced that AS602801, a second generation JNK inhibitor for autoimmune and inflammatory conditions, had moved into Phase 1 clinical trials.

The first small molecule inhibitor of JNK to enter clinical trials was CC-401 (Celgene). A corporate press release dated February 5, 2003 stated CC-401, "successfully completed a Phase I, double-blind, placebo controlled, ascending single intravenous dose study in healthy human volunteers", a statement which while lacking in detail does provide some preliminary evidence that acute toxicity may not be an issue with JNK inhibitors. Neither a chemical structure or a detailed characterization of this compound has been publicly disclosed although its hepatocyte protecting activity in a model of rat liver transplant has been described.[35] In these studies CC-401 was found to be most efficacious when present in the liver prior to reperfusion that follows surgical implantation. This is consistent with the dogma that it is the reperfusion event that leads to rapid induction of JNK activity in hepatocytes.[36] It was also observed that CC-401 inhibited cytoplasmic cytochrome c providing confirmatory evidence that JNK does play a role in the mitochondria dependent apoptosis pathway, at least in settings of oxidative injury to the cell. It is interesting to note that the initial efficacy data for AS601245 and CC-401 have independently focused on the area of cellular apoptosis caused by ischemic or ischemia-reperfusion injury suggesting this may be a promising clinical development route for JNK inhibitors.

The identification and development of JNK inhibitors as therapeutics is a competitive arena and numerous chemical classes of JNK inhibitors have been disclosed in patent applications, mostly from pharmaceutical companies. As much of this work is proprietary it is possible that one or more JNK inhibitors representative of these classes are also in clinical trials. Table 1 gives examples of the various chemical classes that have been claimed. Unfortunately, in the absence of any disclosed information for these compounds, the biologic data for most of these molecules remains confidential.

In this section we have described small molecules that directly inhibit JNK. However there are insightful examples of compounds that affect the JNK pathway by targeting upstream kinases, or ligands of the JNK signaling cascade. CEP-1347 (Cephalon) is an indolocarbazole analog of the prototype kinase inhibitor, K252a.[37] Although originally optimized based on its ability to promote survival of neurons in vitro, it was recognized early on that suppression of JNK activity was likely contributing to the observed cell response.[38] It was later discovered that CEP-1347 was in fact a potent inhibitor of the mixed-lineage kinase family (MLK), which are the MAP3K kinases of the JNK signaling pathway (upstream activators of MKK4/7).[39,40] In line with its therapeutic focus, Cephalon has tested CEP-1347 predominantly in

Table 1. Compilation of JNK inhibitor structures

JNK Inhibitors Cited in Text

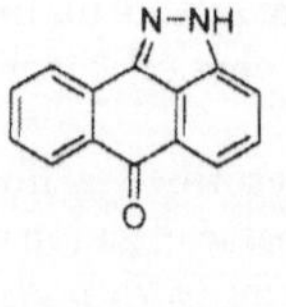

SP600125
Signal Pharmaceuticals
JNK2: ~100 nM

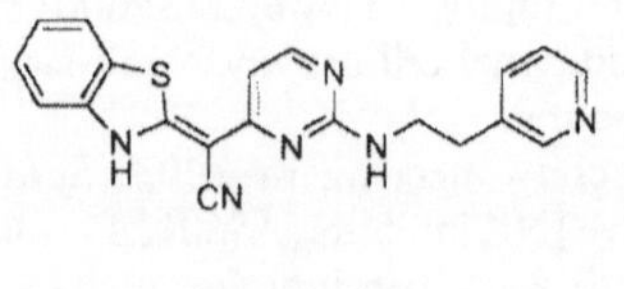

AS-601245
Serono
JNK3: 50 nM

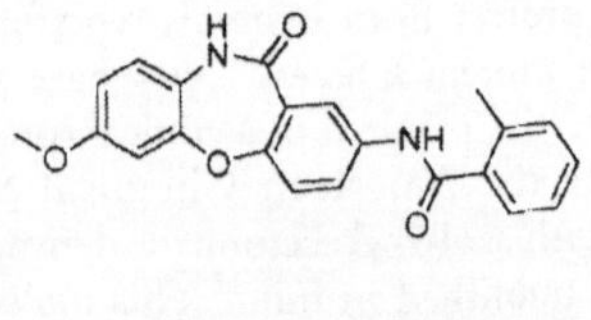

Benzoxazepinones
Amgen
JNK3: 47 nM

JNK Inhibitors in Patent Literature

Pyridines/Pyrimidines

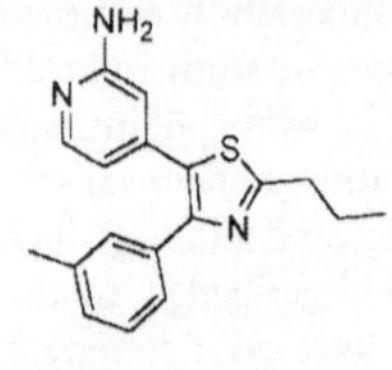

WO 2002062792
Takeda
JNK1: <100 nM

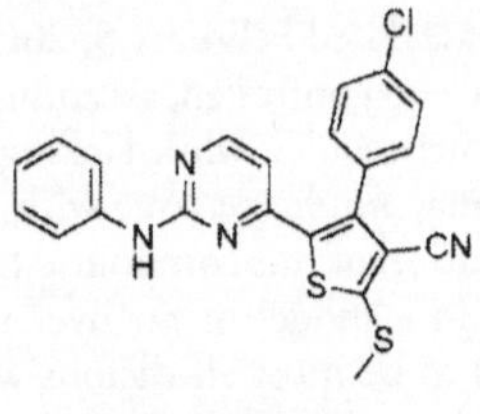

WO 2002083667
Vertex
JNK3: <100 nM

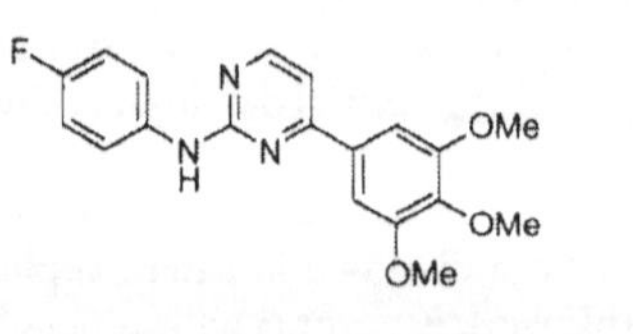

WO 2002079197
Vertex
JNK3: <100 nM

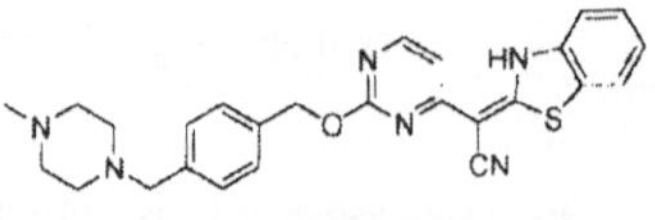

WO 2003091249
Serono
JNK3: <10 μM

Table continued on next page

neurodegenerative conditions. These data include efficacious responses in mouse and monkey MPTP (dopaminergic neuron toxin) models of Parkinson's disease[38,41] and auditory hair loss in guinea pig cochleas after exposure to intense noise.[42] CEP-1347 is currently in a Phase 2/3, double blind, placebo controlled trial with an enrollment of 800 Parkinson's disease patients.

Steroids remain the gold standard therapy for many diseases because of their broad effect on multiple pro-inflammatory mediators. The overlap in steroid and JNK biology in cell assays

Table 1. Continued

Indazoles

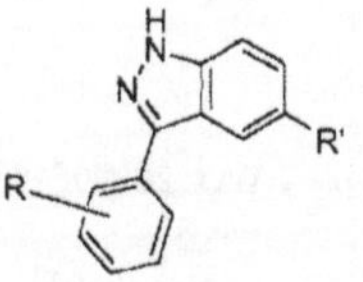

WO 2002010137
Celgene

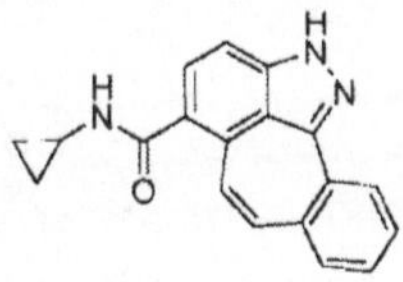

WO 2003072550
Eisai
JNK3: <300 nM

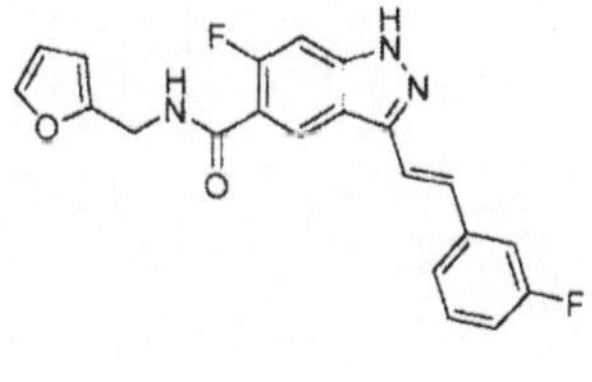

WO 2003101968
Eisai
JNK3: 55 nM

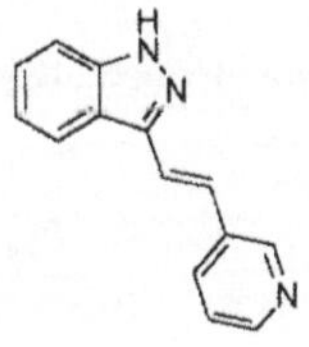

WO 2004050088
Kyowa Hakko

Azaindoles

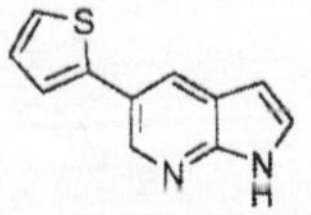

WO 2003082869
Eisai
JNK3: <500 nM

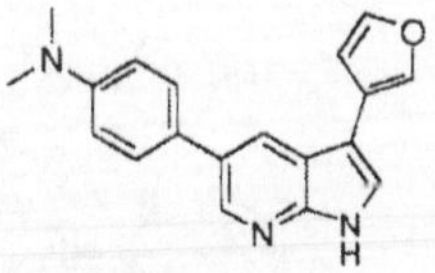

WO 2004078756
Eisai
JNK3: <500 nM

Table continued on next page

has previously been noted.[43,44] Furthermore it is interesting that asthmatics refractory to corticosteroid therapy were observed to have elevated AP-1 and JNK activity in peripheral leukocytes.[45] Multiple mechanisms of action have been proposed for the glucocorticoid receptor (GR) including association and competition with c-Jun/AP-1 at gene promoters. More recently, it has been reported that ligated monomeric GR can bind directly to JNK and that this

Table 1. Continued

Isoquinolinones

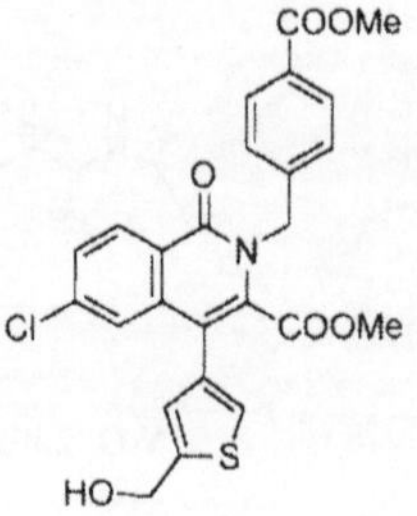

WO 2004054979
Takeda
JNK1: 24 nM

Thienylmethylbenzamides

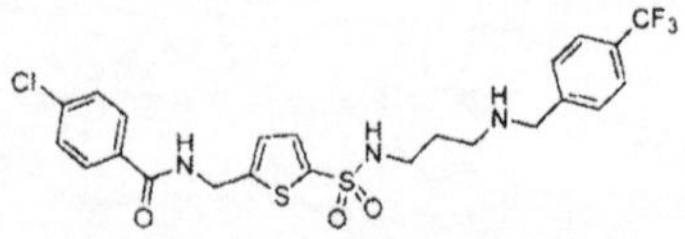

WO 2003010164
Serono
JNK3: 188 nM

Other Structures

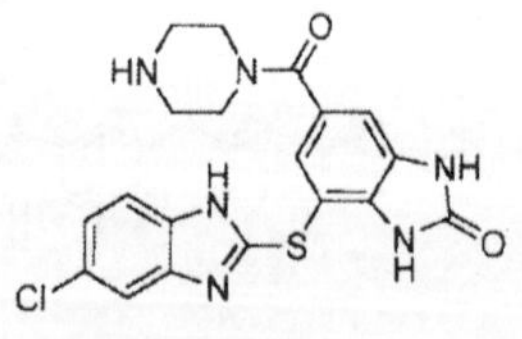

WO 2003051277
AstraZeneca

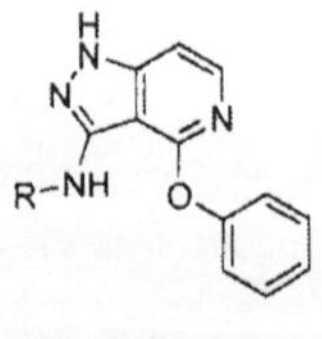

WO 2004113303
AstraZeneca

interaction prevents association of JNK to its upstream activators MKK4/7.[46] Is it possible that this mechanism of action is also shared with the estrogen receptor?[47] This finding suggests that an allosteric site, potentially at the interface of JNK/MKK binding, is therapeutically relevant. Whether this site can be targeted using a small molecule remains to be seen. Furthermore, if the inhibition of JNK signaling by glucocorticoid is physiologically relevant it confirms that

JNK is an important therapeutic target for multiple inflammatory diseases and may even hold as yet unknown advantages over steroid therapy.

Peptide Inhibitors

Small molecule drugs have many pharmacologic and commercial advantages over biotechnology derived alternatives. That the majority of drugs are small molecule inhibitors is mostly due to our historical working knowledge of chemistry compared to biotechnology (vaccines excluded). However, macromolecules such as antibodies, antisense, siRNA or decoy oligonucleotides, and viral-delivered proteins may be as efficacious as the best synthetic compounds. Therefore the potential of such alternative therapeutics should be considered, as technology breakthroughs will certainly improve their clinical and commercial viability. One proof of concept therapeutic reagent specific to JNK is the peptide that encodes a portion of JIP, the JNK interacting protein.[48] Although first described as an inhibitor of JNK based on over-expression studies, JIP (also called IB1) was quickly reclassified as the first of several JNK scaffold proteins.[49,50] Different JNK scaffold proteins probably allow the assembly and specific geographical positioning of unique MLK-MEKK/MKK/JNK enzyme complexes in the cell. Presumably these complexes are also activated by discrete stimuli. The original observation of JIP as an inhibitor of JNK signaling suggested that excess JIP could act as a decoy to sequester JNK and prevent formation of an effective signaling cascade. However, proteins make for poor intracellular therapeutics since they are unable to cross the cell membrane. Bonny and colleagues showed that an 18-20 amino acid peptide within the JIP/JNK interface was as effective a dominant negative as the full-length protein. Addition of an HIV-TAT fusion sequence facilitated membrane translocation in multiple cell types.[51,52] The utility of this peptide was further enhanced by synthesizing the 'retroinverso' version of the peptide, that is, the reverse sequence as D-amino acids. Structurally this approach does not at first appear sensible as the L- and D- versions are distinct molecules. Activity and specificity appear to be partly explained by the observation that it is not the peptide backbone that interacts with JNK, but spatial orientation of critical proline, arginine, and leucine sidechains that are conserved in the retroinverso version.[20,53] This version of the peptide is proteolytically stable and has permitted dosing in animal models of disease including cerebral ischemia where the peptide has exhibited striking neuroprotective effects.[54] Whether this peptide or a peptide mimetic has a clinical and commercial future remains to be seen but it creatively illustrates alternative strategies that could be employed to design novel inhibitors of JNK signaling.[55]

Summary

It has been 15 years since the identification of JNK, and in that time many assumed functions have been confirmed and many novel roles have been discovered. However, the importance of JNK as a critical driver of disease and its potential as a drug target has only been enhanced by these observations. Robust and high-throughput assays are available and we have the atomic structure of the kinase resolved at high resolution. We have achieved selectivity between JNK and closely related MAPK family members and are teasing out the possibility of JNK isoform selectivity to further refine the pharmacology of inhibitors. At least 2 optimized compounds that specifically target JNK are in clinical trials and others that inhibit JNK pathway activity are also under evaluation. The investment within the pharmaceutical industry towards JNK is significant and we will soon learn whether the potential of this innovative target can be effectively realized.

Acknowledgements

We are grateful to all our team members at Celgene, and our colleagues in academia and industry who have, and continue to work towards the successful development of JNK inhibitors as a novel therapy. We thank Elise Sudbeck for generating the image of SP600125 bound to JNK as shown in Figure 1.

References

1. Kyriakis J, Avruch J. pp54 microtubule-associated protein 2 kinase. A novel serine/threonine protein kinase regulated by phosphorylation and stimulated by poly-L-lysine. J Biol Chem 1990; 265(28):17355-17363.
2. Hibi M, Lin A, Minden A et al. Identification of an oncoprotein and UV-responsive protein kinase that binds and protentiates the c-Jun activation domain. Genes Devel 1993; 7:2135-2148.
3. Kyriakis J, Banerjee P, Nikolakaki E et al. The stress-activated protein kinase subfamily of c-Jun kinases. Nature 1994; 369(6476):156-160.
4. Chen C-Y, Gherzi R, Andersen JS et al. Nucleolin and YB-1 are required for JNK-mediated interleukin-2 mRNA stabilization during T-cell activation. Genes Devel 2000; 14(10):1236-1248.
5. Hirosumi J, Tuncman G, Chang L et al. A central role for JNK in obesity and insulin resistance. Nature 2002; 420:333-336.
6. Tournier C, Hess P, Yang DD et al. Requirement of JNK for stress-induced activation of the cytochrome c-mediated death pathway. Science 2000; 288(5467):870-874.
7. Lei K, Davis RJ. JNK phosphorylation of Bim-related members of the Bcl2 family induces Bax-dependent apoptosis. Proc Natl Acad Sci USA 2003; 100(5):2432-2437.
8. Jacobs-Helber SM, Sawyer ST. Jun N-terminal kinase promotes proliferation of immature erythroid cells and erythropoietin-dependent cell lines. Blood 2004; 104(3):696-703.
9. Rincon M, Pedraza-Alva G. JNK and p38 MAP kinases in CD4+ and CD8+ T cells. Immunol Rev 2003; 192, 131-142.
10. Jasper H, Benes V, Schwager C et al. The genomic response of the Drosophila embryo to JNK signaling. Dev Cell 2001; 1(4):579-586.
11. Stronach B, Perrimon N. Activation of the JNK pathway during dorsal closure in Drosophila requires the mixed lineage kinase, slipper. Genes Devel 2002; 16(3):377-387.
12. Cuconati A, White E. Viral homologs of BCL-2: role of apoptosis in the regulation of virus infection. Genes Devel 2002; 16(19):465-2478.
13. Cohen P. Protein kinases—the major drug targets of the twenty-first century? Nat Rev Drug Discov 2002; 1(4):309-315.
14. Murray BW, Bennett BL, Sasaki D. Analysis of pharmacologic inhibitors of c-Jun N-terminal kinase. Methods Enzymol 2001; 332:432-452.
15. Burke J, Pattoli M, Gregor K et al. BMS-345541 is a highly selective inhibitor of I kappa B kinase that binds at an allosteric site of the enzyme and blocks NF-kappa B-dependent transcription in mice. J Biol Chem 2003; 278(3):1450-1456.
16. Dudley D, Pang L, Decker S et al. A synthetic inhibitor of the mitogen-activated protein kinase cascade. Proc Natl Acad Sci USA 1995; 92(17):7686-7689.
17. Fabian M, Biggs III W, Treiber D et al. A small molecule-kinase interaction map for clinical kinase inhibitors. Nat Biotechnol 2005; online.
18. Xie X, Gu Y, Fox T et al. Crystal structure of JNK3: a kinase implicated in neuronal apoptosos. Structure 1998; 6(8):983-991.
19. Scapin G, Patel S, Lisnock J et al. The structure of JNK3 in complex with small molecule inhibitors: structural basis for potency and selectivity. Chem Biol 2003; 10(8):705-712.
20. Heo Y, Kim S, Seo C et al. Structural basis for the selective inhibition of JNK1 by the scaffolding protein JIP1 and SP600125. EMBO J 2004; 23(11):2185-2195.
21. Westwick J, Brenner D. Methods for analyzing c-Jun kinase. Methods Enzymol 1995; 255:342-359.
22. Karin M, Liu Z, Zandi E. AP-1 function and regulation. Curr Opin Cell Biol 1997; 9(2):240-246.
23. Harper S, LoGrasso P. Inhibitors of the JNK signaling pathway. Drugs Future 2001; 26(10):957-973.
24. Bennett BL, Sasaki DT, Murray BW et al. SP600125, an anthrapyrazolone inhibitor of Jun N-terminal kinase. Proc Natl Acad Sci USA 2001; 98(24):13681-13686.
25. Han Z, Boyle DL, Chang L et al. c-Jun N-terminal kinase is required for metalloproteinase expression and joint destruction in inflammatory arthritis. J Clin Invest 2001; 108(1):73-81.
26. Han Z, Chang L, Yamanishi Y et al. Joint damage and inflammation in c-Jun N-terminal kinase 2 knockout mice with passive murine collagen-induced arthritis. Arthritis Rheum 2002; 46(3):818-823.
27. Maeda S, Chang L, Li Z et al. IKKbeta is required for prevention of apoptosis mediated by cell-bound but not by circulating TNFalpha. Immunity 2003; 19(5):725-737.
28. Nath P, Eynott PR, Leung SY et al. Potential role of c-Jun NH2-terminal kinase in allergic airway inflammation and remodelling: effects of SP600125. Eur J Pharmacol 2005; 506(3):273-283.
29. Wang W, Shi L, Xie Y, Ma C, Li W, Su X et al. SP600125, a new JNK inhibitor, protects dopaminergic neurons in the MPTP model of Parkinson's disease. Neurosci Res 2004; 48(2):195-202.

30. Ishii M, Suzuki Y, Takeshita K et al. Inhibition of c-Jun NH2-terminal kinase activity improves ischemia/reperfusion injury in rat lungs. J Immunol 2004; 172(4):2569-2577.
31. Bain J, McLauchlan H, Elliot M et al. The specificities of protein kinase inhibitors: an update. Biochem J 2003; 371(1):199-204.
32. Joiakim A, Mathieu P, Palermo C et al. The Jun N-terminal kinase inhibitor SP600125 is a ligand and antagonist of the aryl hydrocarbon receptor. Drug Metab Dispos 2003; 31(11):1279-1282.
33. Carboni S, Hiver A, Szyndralewiez C et al. AS601245 (1,3-benzothiazol-2-yl (2-[[2-(3-pyridinyl) ethyl] amino]-4 pyrimidinyl) acetonitrile): a c-Jun NH2-terminal protein kinase inhibitor with neuroprotective properties. J Pharmacol Exp Ther 2004; 310(1):25-32.
34. Ferrandi C, Ballerio R, Gaillard P et al. Inhibition of c-Jun N-terminal kinase decreases cardiomyocyte apoptosis and infarct size after myocardial ischemia and reperfusion in anaesthetized rats. Br J Pharmacol 2004; 142(6):953-960.
35. Uehara T, Xi Peng X, Bennett B et al. c-Jun N-terminal kinase mediates hepatic injury after rat liver transplantation. Transplantation 2004; 78(3):324-332.
36. Bradham C, Stachlewitz RF, Gao W et al. Reperfusion after liver transplantation in rats differentially activates the mitogen-activated protein kinases. Hepatology 1997; 25(5):1128-1135.
37. Kaneko M, Saito Y, Saito H et al. Neurotrophic 3,9-bis[(alkylthio)methyl]-and-bis(alkoxymethyl)-K-252a derivatives. J Med Chem 1997; 40(12):1863-1869.
38. Saporito MS, Brown EM, Miller MS et al. CEP-1347/KT-7515, an inhibitor of c-jun N-terminal kinase activation, attenuates the 1-methyl-4-phenyl tetrahydropyridine-mediated loss of nigrostriatal dopaminergic neurons in vivo. J Pharmacol Exp Ther 1999; 288(2):421-427.
39. Maroney AC, Finn JP, Connors TJ et al. Cep-1347 (KT7515), a semisynthetic inhibitor of the mixed lineage kinase family. J Biol Chem 2001; 276(27):25302-25308.
40. Wang L, Besirli C, Johnson Jr E. Mixed-lineage kinases: a target for the prevention of neurodegeneration. Annu Rev Pharmacol Toxicol 2004; 44:451-474.
41. Saporito MS, Hudkins RL, Maroney AC. Discovery of CEP-1347/KT-7515, an inhibitor of the JNK/SAPK pathway for the treatment of neurodegenerative diseases. Prog Med Chem 2002; 40:23-62.
42. Pirvola U. Rescue of hearing, auditory hair cells, and neurons by CEP-1347/KT7515, an inhibitor of c-Jun N-terminal kinase activation. J Neurosci 2000; 20(1):43-50.
43. Swantek J, Cobb M, Geppert T. Jun N-Terminal Kinase/Stress-Activated Protein Kinase (JNK/SAPK) is Required for Lipopolysaccharide Stimulation of Tumor Necrosis Factor Alpha. Mol Cell Biol 1997; 17(11):6274-6282.
44. Bantel H, Schmitz M, Raible A et al. Critical role of NF-kappaB and stress-activated protein kinases in steroid unresponsiveness. FASEB J 2002; 16(13):1832-1834.
45. Sousa A, Lane S, Soh C et al. In vivo resistance to corticosteroids in bronchial asthma is associated with enhanced phosyphorylation of JUN N-terminal kinase and failure of prednisolone to inhibit JUN N-terminal kinase phosphorylation. J Allergy Clin Immunol 1999; 104(3):565-574.
46. Bruna A, Nicolas M, Munoz A et al. Glucocorticoid receptor-JNK interaction mediates inhibition of the JNK pathway by glucocorticoids. EMBO J 2003; 22(22):6035-6044.
47. Srivastava S, Weitzmann M, Cenci S et al. Estrogen decreases TNF gene expression by blocking JNK activity and the resulting production of c-Jun and JunD. J Clin Invest 1999; 104(4):503-513.
48. Dickens M, Rogers J, Cavanagh J et al. A cytoplasmic inhibitor of the JNK signal transduction pathway. Science 1997; 277(5326):693-696.
49. Whitmarsh AJ, Cavanagh J, Tournier C et al. A mammalian scaffold complex that selectively mediates MAP kinase activation. Science 1998; 281(5383):1671-1674.
50. Bonny C, Nicod P, Waeber G. IB1, a JIP-1-related nuclear protein present in insulin-secreting cells. J Biol Chem 1998; 273(4):1843-1846.
51. Bonny C, Oberson A, Negri S et al. Cell-permeable peptide inhibitors of JNK: novel blockers of beta-cell death. Diabetes 2001; 50(1):77-82.
52. Vives E, Brodin P, Lebleu B. A truncated HIV-1 Tat protein basic domain rapidly translocates through the plasma membrane and accumulates in the cell nucleus. J Biol Chem 1997; 272(25):16010-16017.
53. Barr R, Kendrick T, Bogoyevitch M. Identification of the critical features of a small peptide inhibitor of JNK activity. J Biol Chem 2002; 277(13):10987-10997.
54. Borsello T, Clarke P, Hirt L et al. A peptide inhibitor of c-Jun N-terminal kinase protects against excitotoxicity and cerebral ischemia. Nat Med 2003; 9(9):1180-1186.
55. Bogoyevitch M, Boehm I, Oakley A et al. Targeting the JNK MAPK cascade for inhibition: basic science and therapeutic potential. Biochim Biophys Acta 2004; 1697(1-2):89-101.

Index

J

K

L

M

N

O

P

R

S